AF522357

Forest Conservation and Management

Forest Conservation and Management

Nandini Sarkar

RANDOM PUBLICATIONS
NEW DELHI (INDIA)

Forest Conservation and Management

978-93-5111-160-3

Published in 2013 in India by

RANDOM PUBLICATIONS

4376-A/4B, Gali Murari Lal, Ansari Road
New Delhi-110 002
Phone : +9111-43580356, 011-23289044
e-mail: randompublications@hotmail.com

Reprinted 2019

Type Setting by : Friends Media, Delhi-110089
Digitally Printed at: Replika Press Pvt. Ltd.

Preface

This book is especially that will help us to correct the harmful effects of strategies and policies adopted in the past and pursue management practices that will safeguard the ecological balance of the region and provide livelihood security for the indigenous people who are dependent on forests.

Forests provide essential social, environmental and economic value to people around the world. However, the economic and practical benefits gained from extracting raw materials for goods from forests are often at odds with the social and especially the ecological services forests provide. Conservation is necessary to ensure forests are able to provide these essential services for generations to come, but management of forest resources is also required to meet global demand for goods and employment.

Forest management expenditures, which are made in the expectation of creating benefits at some time in the future, are best viewed as investments. Although some foresters are reluctant to take this perspective, there are compelling reasons for doing so. This opens up opportunities for using standard investment analysis techniques to rank projects competing for scarce funding. The use of these techniques lets forest managers evaluate alternative strategies and select the one which will contribute the most to the achievement of management goals. This technical note is intended to provide forest resource managers with an introduction to benefit-cost analysis and its potential use in forest management. We present a basic introduction to some important concepts of benefit-cost analysis, then discusses the nature and valuation of benefits and costs commonly found in forestry

This book is expected to benefit the researchers, general readers and policy makers for policy implication.

—Author

Contents

1

Forest Conservation and Management

Forests provide essential social, environmental and economic value to people around the world. However, the economic and practical benefits gained from extracting raw materials for goods from forests are often at odds with the social and especially the ecological services forests provide. Conservation is necessary to ensure forests are able to provide these essential services for generations to come, but management of forest resources is also required to meet global demand for goods and employment.

There are three main approaches to forest conservation: protection, sustainable forest management and restoration. A forest management plan can include components of all three.

FOREST PROTECTION

Protection is a regulatory approach that involves creating "protected areas" that restrict potentially damaging activities. According to the International Union for Conservation of Nature (IUCN), a protected area is: "a clearly defined geographical space, recognized, dedicated and managed, through legal or other effective means, to achieve the long-term conservation of nature with associated ecosystem services and cultural values". There are six different IUCN protected area categories, each defined to protect and manage an area for a range of purposes. For instance, areas are protected and managed for conservation, science, recreation and sustainable use. Protected areas cover 13.5% of the world's forests. In Canada, more than 31 million hectares (8%) of forest and other wooded land are within protected areas.

SUSTAINABLE FOREST MANAGEMENT

Since little of the world's forests are protected, conservation often occurs alongside the production of forest materials. This is where Sustainable Forest Management (SFM) comes into play – it attempts to ensure a balance between the objectives of conservation and production. SFM can be implemented in different ways, but generally it means taking care of forests in a way and at a rate that allows them to maintain their long-term health and fulfill ecological, economic, social and cultural functions now and in the future.

One way to promote SFM is through the marketplace. For example, companies have more of an incentive to practice SFM if they know consumers prefer responsibly harvested products. One initiative that addresses this issue is called third-party certification. To become certified, a company must have their on-the-ground forest operations audited by an independent third party, which compares their practices against pre-determined standards. Certification schemes also follow goods from the raw material stage through to the finished product (*e.g.* paper) to make sure the entire "chain of custody" complies with set standards.

Certification allows forest companies to provide customers with a guarantee that the products they purchase come from sustainably managed forests. However, there are three different certification standards used in Canada and some are considered more stringent than others. The Canadian Standards Association (CSA), the Sustainable Forest Initiative (SFI) and the Forest Stewardship Council (FSC) are the certification schemes used in Canada. By June 2009, almost 146 million hectares of Canada's forests were certified under one of the three systems – more than any other country. While only 10% of the world's forests are certified,about 40% of this amount is in Canada.

The government of Canada does not endorse one type of certification over another. However, many environmental organizations argue that CSA and SFI are not as rigorous as FSC, and that FSC is the only credible system. Over two thirds of Canada's forests are certified through CSA and SFI. You can do your own research and check labels on the products you buy to determine how forest-friendly they are.

Another market-based initiative is Payment for Environmental Services (PES). PES schemes encourage sustainable management and maintenance of forest ecosystems by providing incentives (in the form of payments or subsidies) to landowners from the beneficiaries of the services supplied by these ecosystems. If sustainable practices are as profitable as clearing land for agriculture, for example, those with land upstream are more motivated to ensure a clean and reliable supply of water for those downstream.

FOREST RESTORATION

The third approach to forest conservation is restoration, which involves repairing a forest that has been degraded. Specifically, restoration aims to recreate the structure, productivity and species diversity of the original forest so that in time, the ecological processes of the restored forest will match those of the original forest, where possible. IUCN and World Wildlife Fund (WWF) promote a method called Forest Landscape Restoration (FLR), which goes beyond simple tree planting. FLR is a planned process that brings together different stakeholders to identify, negotiate and implement practices that aim to restore ecological integrity and improve human welfare in forest areas that have been deforested or damaged.

Re-planting trees also helps to offset deforestation. The United Nations Environment Programme (UNEP) has created the Billion Tree Campaign, which aims to plant at least one billion trees every year. Over the past decade, 130 million hectares of forest have been lost. To recover this, we would have to plant 14 billion trees every year for the next ten years!

OTHER STRATEGIES

Planted forests (as opposed to naturally regenerating forests)—which account for about 7% of global forest area—are another type of forest management, but they are controversial. Productive planted forests provide significant economic benefits as they are the main global source of wood. Protective planted forests can supply many of the same ecological services as traditional forests. However, planted forests do not provide nearly the same range of biodiversity or aesthetic quality as old-growth forests and their critics claim they have negative consequences for the rural poor, indigenous peoples, migrant and landless workers, women and other disadvantaged groups, and for the environment. Agroforestry is another type of planted forest. It involves the intentional integration of trees into agricultural landscapes. Agroforestry practices offer a variety of environmental services, such as maintaining watershed functions, retaining carbon and supporting the conservation of biodiversity. Community forestry is an approach where local people are directly involved in decisions regarding forest resources. A community forest is any forestry operation managed by a local government, community group, Aboriginal or community-held corporation for the benefit of the entire community. In this type of system, the local community has a vested interest in protecting the environment in which they live. Community forestry has existed for centuries in Brazil, Indonesia, Thailand, Tanzania and India.

SIMPLE METHODS OF FOREST CONSERVATION

For the conservation of forests, following steps can be taken:

(a) Conservation of forest is a national problem so it must be tackled with perfect coordination between forest department and other departments.
(b) People's participation in the conservation of forests is of vital importance. So, we must get them involved in this national task.
(c) The cutting of trees in the forests must be stopped at all costs.
(d) Afforestation or special programmes like Van Mahotsava should be launched on grand scale.
(e) Celebrations of all functions, festivals should precede with tree-plantation.
(f) Cutting of timber and other forest produce should be restricted.
(g) Grasslands should be regenerated.
(h) Forest conservation Act 1980 should be strictly implemented to check deforestation.
(i) Several centres of excellence have been setup and awards should be instituted.

FORESTRY CONSERVATION MANAGEMENT PRACTICES

Forestry Conservation Management Practices (CMPs) are specific, science-based guidelines for conservation of rare species during forest harvesting. CMPs are somewhat analogous to Forestry Best Management Practices (BMPs), except whereas BMPs focus mainly on protection of water resources, CMPs specialize in protection of rare wildlife. The primary objective of CMPs is to guide harvesting activities such that rare species listed under the Massachusetts Endangered Species Act (MESA) are not impacted in a way that jeopardizes long-term viability of local populations. CMPs first identify and describe potential impacts of forest harvesting to state-listed species, whether impacts may be direct (*e.g.*, physical injury or death of individual animals) or indirect (*e.g.*, alteration of habitat in a way that reduces overall reproductive success of a local population). Then, CMPs provide specific guidelines to avoid or minimize impacts that would be considered negative or potentially detrimental to a local population; the guidelines are based on scientific knowledge of the habitat requirements, reproductive strategy, dispersal ability, survivorship, and other ecological factors that influence population dynamics of the species. CMPs are based on the principle that protection and enhancement of the long-term viability of local populations is prerequisite to statewide recovery of imperiled wildlife.

Simultaneous with the objective of protecting local populations of state-listed species, CMPs aim to maintain adequate opportunity for sustainable management of timber products in Massachusetts. To this end, CMPs tend to focus forest harvesting restrictions on the critical areas within known habitat of state-listed species, thereby allowing timber management to proceed with fewer restrictions over as large an area as possible. This strategy is based, in part, on recognition that forest harvesting typically results in temporary habitat change or sometimes even habitat improvement rather than permanent habitat loss. Thus, the CMP strategy is designed to help maximize the protection of state-listed species and the ability of Massachusetts landowners to manage their forests for timber and other wood products.

MANAGEMENT OF FORESTLAND IN MASSACHUSETTS

Management of forestland in Massachusetts involves multiple stakeholders with multiple objectives, so communication among stakeholders and development of management standards are essential to guiding compatible and sustainable uses of the land. State-owned forestland in Massachusetts received Forest Stewardship Council certification in May 2004, and this certification requires that there be coordination of forest management among the three state forest management agencies (Division of Water Supply Protection [DWSP], Bureau of Forestry [BOF], and Division of Fisheries and Wildlife [DFW]). In addition, there is a need for state forestland managers to work closely with the Natural Heritage and Endangered Species Programme (NHESP) to develop and implement a formal and consistent process for determining appropriate mitigation of potentially adverse impacts of forest harvesting to state-listed species on both state and privately-owned forestlands. In light of these goals, CMPs are needed to ensure adequate protection of state-listed species during forest-harvesting operations. Without CMPs, a landowner and/or forester must develop a Forest Cutting Plan without having prior knowledge of possible harvest restrictions required by the NHESP to mitigate potential impacts of forest harvesting to state-listed species. Under these circumstances, activities proposed in the Forest Cutting Plan must often be modified after the plan is reviewed by the NHESP, which tends to result in delay of the harvest, changes in the amount and distribution of timber to be harvested, or changes in the number, type, or location of stream and wetland crossings. However, CMPs shall make potential harvest restrictions available to the landowner and forester before a Forest Cutting Plan is developed. This gives the landowner and forester an opportunity to develop a plan that is more likely to be approved by the NHESP without need for modification, thereby saving the landowner time and money. With proper planning and attention to detail, use of CMPs shall maximize the efficiency of the cutting-plan filing and review process and enable landowners to better understand how forestry and conservation of rare wildlife can be integrated as compatible uses of Massachusetts forestland.

Benefits of CMPs

CMPs benefit both state-listed species and forest management by (1) being based on scientific knowledge of the ecological factors influencing population dynamics of state-listed species, thereby enabling focused protection of key habitats and other resources, (2) improving the predictability and consistency of review of Forest Cutting Plans by the NHESP, and (3) providing detailed guidance to landowners and foresters for developing Forest Cutting Plans that have high probability of being approved by the NHESP without modification. An additional benefit, which is a byproduct of the first listed above, is that time-of-year harvesting restrictions mandated by CMPs tend to apply to smaller areas than those to which restrictions applied in the past. In this sense, intensive conservation measures are concentrated, which should be more effective in achieving conservation objectives than a strategy by which lighter measures are applied over a broader area. Furthermore, this strategy affords the landowner greater liberty in managing a larger proportion of a given tract of forest.

Process for Developing CMPs

The NHESP is the primary partner responsible for development of CMPs. However, staff of the Natural Resources and Environmental Conservation (NREC) Extension Programme at the University of Massachusetts Amherst and foresters of the DWSP, BOF, and DFW collaborate with the NHESP in this effort. This group of scientists and land managers meets regularly to work together on every aspect of the CMP development process. The first step in developing CMPs is to determine which state-listed species should be addressed. This decision is based largely upon the frequency with which forest harvests are planned in habitat of each state-listed species, whereupon species that most frequently coincide with harvests are considered top candidates for development of CMPs. A second factor is the degree to which protection of each state-listed species is expected to impact the logistics of forest harvesting, whereupon species that are likely to require highly protective measures are considered top candidates. A third factor, which often is related to the second, is the sensitivity of a state-listed species or its habitat to forest harvesting, whereupon species that are highly sensitive are considered top candidates. When multiple species have similar ecological requirements and conservation strategies, groups of species may be lumped into a single CMP document (*e.g.*, Blue-spotted Salamander, Jefferson Salamander, and Marbled Salamander are lumped together in a "Mole Salamander" CMP document).

When there is agreement between the NHESP and its collaborators about which state-listed species to address in a CMP document, the NHESP conducts a thorough search of scientific literature pertaining to the ecology of the species of interest. The information is then consolidated into a species account that includes a physical description of the species and a discussion of its range, distribution, habitat requirements, seasonal movements, reproductive strategy, survivorship, and other concepts associated with its life history. Additionally, the account includes a discussion of known and potential threats to population viability, including those posed by forest harvesting.

The NHESP and its collaborators then develop the actual management practices that scientific evidence suggests is necessary for successful conservation of the species. These management practices are specific to forestry-related activities and, like Forestry BMPs, are split into two categories: required management practices and recommended management practices. Required management practices (denoted with "R" in the CMP document) are those which should be considered mandatory during most harvests, barring special circumstances. Recommended management practices (denoted with "G" in the CMP document) are optional guidelines that are expected to provide additional benefits to the species of interest.

METHODS FOR CONSERVING FOREST GENETIC RESOURCES

A wide range of methods, from protected reserves to intensive management of breeding populations for production systems, can be used to conserve FGR. The choice of methods depends on available genetic material, selected time scale and specified aims. The method selected and the subsequent implementation of a conservation strategy also depend on the availability of both human and financial resources. The two most commonly considered methods for conserving FGR are *in situ* and *ex situ* conservation. The term *in situ* refers to the continued maintenance of tree populations at their natural sites, in the environment to which they have adapted. *Ex situ* conservation takes place outside the natural habitat of a tree species, and may consist of activities such as establishing live collections or *ex situ* conservation stands, or storing seeds, pollen or tissue.

FGR can also be conserved and maintained by using tree species in forestry or other land-use systems such as agroforestry. It is likely, however, that forest management interventions reduce genetic variation in tree populations. Not even carefully planned and implemented forest management activities, therefore, can replace more active conservation measures with clearly specified objectives. The same caveat also applies to tree domestication and the use of trees in agroforestry systems.

IN SITU CONSERVATION

There is a general consensus among scientists and practitioners that no single conservation method is adequate, and that different methods should be applied in a complementary manner. *In situ* conservation, however, has a number of benefits and so often forms the basis of conservation programmes. It allows evolutionary processes to be maintained, including the adaptation of tree populations to changing environmental conditions. This is particularly important for breeding programmes, since future human needs and environmental conditions are difficult to predict. The reproductive biology and overall survival of many tropical forest species are dependent on complex ecological interactions. Thus long-term genetic conservation of such species is difficult, if not impossible, without *in situ* conservation. In addition, recalcitrant seed behaviour (*i.e.* inability to tolerate desiccation) also complicates long-term *ex situ* conservation efforts for many tropical tree species. *In situ*conservation can also contribute to the conservation of biological diversity at higher levels, *i.e.* species and ecosystems.

Protected areas have often been established on the basis of ecosystem or species conservation, rather than gene conservation. Thus the design of *in situ* conservation programmes has been considered primitive. In tropical forests in particular, the complexity of interactions and lack of scientific information have hampered the development of *in situ* conservation strategies. Uncertainty surrounds the adequate size of *in situ* conservation areas, the number of individuals to be included, how to select locations, and how genetic variation is distributed within selected areas. Undisturbed tropical forest ecosystems are often taken as a starting point when planning *in situ* conservation programmes. Today, however, rural landscapes are a mosaic of disturbed and less disturbed patches of forest, ranging from seemingly natural and secondary forests to seriously degraded forests and other wooded fragments. The most obvious genetic effects of fragmentation are a loss of genetic diversity at both population and species levels, changes in the genetic structure of a population and increased inbreeding. Clearly, *in situ* efforts cannot always rely on intact natural forests, and so it is essential to understand how the processes of genetic drift, gene flow, selection and mating affect genetic diversity in fragmented forests.

IPGRI's partners at the Universities of Costa Rica, Alberta and Massachusetts have been studying the effects of fragmentation on the genetic diversity of *Enterolobium cyclocarpum* in Costa Rica. Studies of the reproductive biology of *E. cyclocarpum*, a tropical dry forest species, in continuous forests and forest fragments in pastures indicate that pasture trees are less likely to receive pollen and set fruit, and that their fruits bear less seed, than trees located in continuous forests. Also, outcrossing rates within the two groups of trees have been found to be similar, but progeny vigour among seedlings in continuous forest is higher than in pastures. These results have immediate implications for the conservation and management of *E. cyclocarpum* and other species in similar habitats. Trees in pastures can aid the movement of pollinators between intact forest fragments and so contribute to gene flow and maintenance of genetic diversity among forest fragments. However, because seedlings in pastures have less vigour, seeds from these areas cannot be recommended for establishing plantations or rehabilitating degraded natural forests.

In southern India, sandal (*Santalum album*) forests have long been exposed to selective logging, poaching and changes in land use. The effects of disturbance on the genetic diversity of sandal have been investigated by the University of Boston, together with the Ashoka Trust for Research in Ecology and the University of Agricultural Sciences in Bangalore. Results from two protected areas indicate that genetic diversity of sandal is highest in the core undisturbed park zone, whereas allelic diversity is reduced in the outer buffer and disturbed zones. These findings suggest that present protection measures are adequate and that excessive logging will cause genetic deterioration in natural sandal populations. Allelic diversity, however, is similar in the two disturbed zones, which suggests that, contrary to general assumptions, no significant genetic segregation has resulted from different degrees of disturbance. This finding can be attributed to the fact that sandal is insect-pollinated and animal-dispersed, *i.e.* substantial gene flow still occurs despite varying degrees of disturbance.

EX SITU CONSERVATION

In situ and *ex situ* conservation should be used in a complementary fashion to conserve FGR. Both are an integral part of the conservation process, and both can only be effective after genetic diversity has been located and conservation priorities have been set. The main purpose of *ex situ* conservation is to capture and maintain a representative sample of the existing genetic diversity of a species. For highly endangered tree species, *ex situ* conservation may be the only approach in the short to medium term. The main pitfalls of collecting germplasm samples for *ex situ* conservation are: i) limited coverage of genetic variation; ii) biases in the collected plant material; and iii) samples that are too large to deal with. Because *ex situ* conservation is more costly than *in situ*conservation, it is particularly important that sampling of populations and germplasm within populations is given special attention to maximize the use of limited financial and human resources.

The traditional approach to *ex situ* conservation of FGR is to establish conservation stands of a species outside its native habitat to facilitate gene management. However, long-term maintenance of the collected genetic variation in *ex situ* stands tends to be complicated by genetic drift and potential contamination by external gene flow. Another commonly used method of *ex situ* conservation is to store seeds collected from a range of natural populations. In the case of many tropical tree species, however, this method is limited by recalcitrant seed behaviour. More than 70% of commercially valuable tropical tree species are estimated to have recalcitrant or intermediate seeds. In recent years, therefore, considerable effort has been expended in developing *in vitro* techniques for *ex situ* conservation of recalcitrant tree species. A notable example is cryopreservation. Cryopreservation, the storage of cells or tissue at ultra-low temperatures, offers great opportunities for longer-term *ex situ* gene preservation. Its applicability, however, has been limited by difficulties in identifying protocols to reduce the water content of tissue before freezing. Tissue water content must be reduced to avoid lethal ice formation in cells. Different results from different laboratories engaged in drying the same species of seed are an additional obstacle. Although it is likely that further research will ease these problems, the main application of cryopreservation seems to be in supporting tree improvement programmes and conserving biotechnically derived germplasm, rather than as a widely applied *ex situ* conservation method to support *in situ* conservation of FGR.

Micropropagation, the use of tissue and organ cultures for organogenesis and somatic embryogenesis, can also be applied to maintain genetic variation in germplasm collected for *ex situ* conservation. Practical propagation applications are already available for several tropical trees, based on shoot tissue culture in broadleaved tree species and on somatic embryogenesis in conifers. Micropropagation does not necessarily require expensive laboratory facilities, and can therefore be used in less-developed conditions. Maintaining juvenile material, however, can be problematic. At present, the technology used in somatic embryogenesis is in most cases too expensive for cost-effective *ex situ* conservation.

Since 1997, the Danida Forest Seed Centre (DFSC) and IPGRI have cooperated in enhancing *ex situ* conservation methods for recalcitrant tropical forest trees. Project activities have focused on determining whether seeds are recalcitrant, intermediate or orthodox, and have included seed development research to assess optimal conditions for seed collecting, germination and storage. An international Forest Tree Seed Research Network has been established under the project to facilitate information exchange among scientists developing protocols for collecting, handling, testing and screening of tolerance to desiccation and optimal storage conditions. At present more than 20 countries worldwide are involved in the project. More than 50 tropical forest species have been screened so far. Activitics also include publications and training workshops to increase research capacity, particularly in developing countries.

CONSERVATION OF FOREST GENETIC RESOURCES

Although the goal of conserving forest genetic resources can be simply stated, its implementation can be very complex and expensive. With thousands of tree species distributed among several local populations (interbreeding groups of individuals), each with thousands of variable genetic loci, priorities should be set first at the species level; only then can we assign priorities among populations. It is important that data on species that are of significance to conserve, and the levels of threat to them, be collated with in situ and ex situ management approaches in mind. In practical terms, national programmes need ways of establishing priorities for conservation that take into account the large potential number of species for which they may be responsible. Sometimes the focus may be on species, because of their charismatic appeal. Alternatively, the focus may have to be on perceived threats resulting from economic values or ecological traits: for example, species that have low population densities, highly specialized pollination patterns or particular seed germination mechanisms. Baseline information on the status of genetic diversity, a rating of species potential value, an evaluation of the threats and the potential for conservation management are some of the necessary steps for priority-setting. The outcome can be a ranking of priorities for management or a classification of species into priority groups.

In general, an effective species conservation programme needs to take into account the whole range of geographic distribution of a species, as well as the species metapopulation structure. Without this information, one cannot claim that the genetic diversity of the target species is conserved overall. Most national conservation programmes for forest genetic resources must therefore deal with the conservation of locally adapted populations. In an ideal case, the geographic distribution of a species would be listed and mapped, the type and extent of threats to particular populations would be known, and methods of conservation and management would be well established. It would then be possible to evaluate the impacts of different threats, the costs and the effectiveness of different management options in minimizing the associated impacts. Priorities could then be established, based on the evaluation of those resources for economic or ecological payoffs. However, we rarely have such complete or detailed knowledge. Many tree species that are known to be under threat of extinction are not included in conservation programmes. In some instances, the genetic resources may be well conserved within protected areas, but these might represent just a small fraction of the overall genetic variation of the species. Patterns of genetic variation may often be cryptic and reductions may portend demographic collapse as either a causal or an associated agent. Therefore, demographic data, if available, is not sufficient, nor are listings of species in parks sufficient for planning conservation. However, collating such data and testing for genetic patterns of variation can provide at least initial indicators of the state of the genetic resource.

Targeting species and regions for more detailed genetic surveys can then provide information on priority species for conservation. These authors also suggest two levels of targeting; one for research and risk assessments, and the second for management of specific risks. Large-scale data, such as are available from geographic information systems (GIS) analyses, can provide some indications of demographic threats, and hence multiple sources of data to be collated and analysed simultaneously. In the first targeting efforts of the former IBPGR (International Board of Plant Genetic Resources, currently International Plant Genetic Resources Institute, IPGRI), the priority lists of crop species that guided collection efforts were created largely on the basis of likelihoods of demographic extinction and thus on risks of genetic erosion, made possible by a broad global sharing of common objectives and information from crop species in agricultural systems. Unfortunately, there is no such consensus for forest trees and hence more explicit statements of models and factors are needed at an international level, as well as for any country trying to establish its own priorities.

In most conservation programmes, people must make decisions about management actions, even with limited access to research data to support their decisions. In these cases, they must also decide which characteristics of the organisms involved are least known but, if studied, are most likely to make a difference in how the organisms are managed. In this sense, we are interested not so much in descriptive biology as in the biology of genetically and ecologically critical functions, such as adaptation to changing environments and potential for evolution and artificial selection.

CONVERSION OF FORESTS TO OTHER LAND USES

Forests can be lost either because forest resources and trees are not regarded as being of economic importance, or because of a policy framework that makes it possible to replace forests with other land uses. Often this is based on short-term maximization of economic returns and lack of supportive forest policies based on good understanding of the potential of forests as sources of income and products for local and regional markets and their associated services for other sectors of the economy.

CONSERVE: ECOSYSTEMS, SPECIES OR GENES

The first step in genetic conservation is to specify the objectives of the conservation programme. This is of the utmost importance, since it is possible to conserve ecosystem properties and still lose species entities. It is also possible to conserve a species and still lose genetically distinct populations, and therefore, genes that may be of value in disease and pest resistance, and in future adaptation. They could also be important in species deployment through breeding programmes, if that becomes a need or necessity. Many forest genera and species around the world provide goods and services, such as timber, wood, food, fodder, environmental stabilization, shade, shelter, and cultural and spiritual values. However, fewer than 1 000 tree species have been systematically tested for their present-day utility, and less than 100 are the subject of intensive genetic research programmes. Evidently, therefore, many different forest species are being used in situ to provide important goods and services, without any active genetic management. Worldwide, the conservation of forest genetic resources has as its overall objective the maintenance of genetic diversity in the thousands of tree species of known or potential socioeconomic and environmental importance. Moreover, the levels and distribution of genetic variation in any given species are expected to be in a process of constant natural change resulting from the main forces of evolution. Therefore, the central concern of conservation should be the evolutionary processes which promote and maintain genetic diversity, and not the endeavour to preserve the present distribution of variation as an end in itself.

ASSESSING SPECIES' PRIORITIES FOR CONSERVATION ACTION

Within any given country or local area there may be divergent opinions on priorities among tree species. Forestry departments are likely to have a somewhat different emphasis and priorities from those of local forest dwellers and users, which can be different again from those of farmers and various other users of trees. It is apparent that in situ conservation programmes will be more successful if they target species of direct interest, use or concern to the land management authority and/or landowner(s): this will have major implications for the planning of in situ conservation programmes. A participatory rural appraisal approach can be useful for helping local communities whose land is communally owned to better identify their priority tree genetic resources and to develop appropriate in situ conservation responses.

It is also important to consider the case for species and provenances which are of major economic importance when planted as exotics, but currently of much less significance in their native range and habitats; an example is Pinus radiata from the south-western USA and Mexico. In such cases it is not unreasonable to expect that the likely beneficiaries of in situ conservation should contribute, financially or otherwise, to conservation. Given that there will be limited financial resources available for specific conservation programmes for forest genetic resources, it is necessary to consider which of the priority species are also in most need of, or warrant, conservation interventions and actions. This can be conveniently undertaken for different species by comparing the extent of the resource (level of genetic diversity or intraspecific variation) with the vulnerability or threats to the populations and/or ecosystems of which they are a part.

In Situ Conservation Atrategies

In the case of non-domesticated species, in situ conservation is probably the most important strategy and sometimes the only viable approach. In the tropics, where extinction rates of species are high because of land-use changes, setting conservation priorities is critical. This is particularly evident in developing countries, where resources allocated for conservation are scarce and baseline information on species distribution and richness data are lacking. In a world of scarce resources, one approach to priority-setting is through networking activities, with initiatives involving multiple countries and stakeholders.

In situ conservation is usually the preferred conservation strategy for most wild plant species, including some of the wild relatives of crop species, because, it allows the populations of interest to continue to be exposed to evolutionary processes. Alternatively, for many domesticated species (crop and livestock), on-farm conservation of traditional varieties is now widely supported as an important practice for conservation of genetic diversity.

Molecular genetic studies, carried out on many forest tree species around the world, are contributing to a better understanding of patterns of variation to support the development of improved management practices, and to monitor changes of species turnover in time and in space. In some situations, these priorities could be refined by the use of new tools, such as molecular markers and modelling simulations.

Integrating GIS tools with molecular research will improve our knowledge of landscape patterns of genetic diversity of species distribution, and help develop resource management plans. For example, in the Western Ghats, India, these two approaches are being used in combination to detect areas with high diversity (intra- and interspecific) and to set priorities for conservation. Molecular markers can thus assist in the conservation of tree species and may allow national programmes to reflect biodiversity patterns in their own management plans. Molecular methods can also help to identify differences between local and non-local provenances and genotypes, identify where diversity is being lost, and facilitate the introduction of new diversity for integrated conservation and breeding programmes. This can also provide for a better management of intraspecific genetic variation.

Evolutionary Conservation

Evolutionary conservation activities are characterized by programmes where the trees produce progeny in successive generations: genes are generally 'conserved', but genotypes are not. Natural selection takes place among trees with new allelic combinations that either favour or disfavour different genotypes. This process ensures that gene frequencies will change in the population: alleles with positive influence on fitness will increase, and alleles associated with low fitness will decrease. If the population size is sufficient, neutral genes should, in general, be maintained, but some genes will inevitably be lost by genetic drift; new genetic variation will arise by mutation after several generations. Human interventions (if any) are designed to facilitate moderate genetic processes rather than to avoid them. Genetic variation between populations is generally maintained when they are growing in different environments, and is even expected to increase over time.

A typical example of a conservation population capable of evolutionary processes is a protected area in a natural forest. In a protected area, the species occupies its natural habitat (it is said to be in situ conserved), typically with a wide range of other species. Natural selection for general fitness is therefore largely related to competition among species, as well as to adaptation within species to current and future environmental conditions. However, evolutionary conservation can also take place in a planted stand, if natural selection is allowed to work, and if the planted trees are regenerated from seed, rather than by vegetative techniques, for the next generation. In such programmes, plantations will preferably be established and managed in ways that mimic the natural processes that will support natural selection. Of course, in most situations the mixture of species (if any) is largely artificial, and the selective forces may therefore favour different genes than would be the case in true in situ conservation. However, this reflects the fact that selection and fitness always depend on the degree of human influence in any ecosystem. Directional selection in favour of commercial traits—including characters such as good stem form or ease of establishment in plantations—is typically avoided in strict evolutionary conservation programmes, but of course this again depends upon the local objectives of the programme.

2

Indian Forestry

GLOBAL DISCOURSE ON FORESTRY

"Liberty and forest laws are incompatible," remarked an English country vicar, speaking on behalf of villagers shut out of woodland reserved for the exclusive use of the king, in 1720. The history of state forestry is indeed a history of social conflict. In monarchies and in democracies, in metropolitan Europe as well as in colonial South Asia, the state management of forests has met bitter and continuous opposition. On the one side are the professional foresters who believe that timber production can be ensured only through the exclusion of humans and their animals from wooded areas; on the other, the peasants, pastoralists, charcoal ironmakers, basketweavers, and other such groups for whom access to forests and forest resources is crucial to economic survival. Environmentalists have added to the criticisms of these latter groups, charging foresters with simplifying complex ecosytems in the direction of commercially valuable but biologically impoverished monocultures.

These contending parties have battled for more than two hundred years. In continental Europe, the eighteenth and nineteenth centuries were peppered with social protest movements against the state management of forests. These protests inspired, among other things, Karl Marx's first political writings and a memorable novel by Honore de Balzac capturing peasant hostility to forest officials. When the European model of strict state control over forests was exported to the colonies, the disaffected peasants and tribals responded with arson and violence. Movements over forest rights were a recurring phenomena in colonies ruled by the British, the Dutch, and the French. The conflicts persisted when the post-colonial governments of countries such as Malaysia and Indonesia followed the authoritarian model of forest management inherited from the colonizer.

In recent decades, however, the global discourse on forestry has moved towards a more accommodationist perspective. Foresters and peasant protesters now seem to talk to, rather than talk past, each other. A willingness to listen to and at least partially incorporate the other point of view has replaced the rigid and uncompromising attitude of the past. Within the forestry profession itself, skeptics doubt the contemporary relevance of the custodial and policing approaches previously followed. A system of natural resource management crafted in absolutist and colonialist times clearly needed to be seriously modified or even overthrown.4 Social activists and community leaders have, meanwhile, moved from demanding a total state withdrawal from forest areas to asking governments to more seriously and sympathetically consider the rights of forest-dependent communities.

With this move from conflict to collaboration have come shifts in the language of forestry itself. Terms such as "scientific forestry" and "rational land management," euphemisms for state control and commercial timber production, are being rapidly replaced by sweet-sounding phrases such as "community management," "participatory development," and "joint forest management." While these terms have come into vogue in the last two decades, they have, in fact, a very long geneaology. From the beginnings of state forestry, there have been serious attempts to democratize the regimes of resource management. Both dissidents within the bureaucracy as well as intellectual activists outside it tried hard to make the state respond more sensitively to the just claims of local communities. The ongoing programmes of joint forest management in India can draw legitimacy and sustenance from a struggle that is at least a century old.

THE LAW AND THE PROTESTS

The crucial watershed in the history of Indian forestry is undoubtedly the building of the railway network. In a famous minute of 1853, the governor general of India, Lord Dalhousie, wrote of how railway construction was both the means for creating a market for British goods and the outlet for British capital seeking profitable avenues for investment. Thus between 1853 and 1910 more than eighty thousand kilometers of track were laid in the subcontinent.6 The early years of railway expansion witnessed a savage assault on the forests of India. Great chunks of forest were destroyed to meet the demand for railway sleepers (over a million of which were required annually). The sal forests of Garhwal and Kumaun, for example, were "felled in even to desolation." "Thousands of trees were felled which were never removed, nor was their removal possible."

This depradation brought home most forcefully the fact that India's forests were not inexhaustible. At this time the British were unquestionably the world leaders in deforestation, having burnt or felled hundreds of thousands of acres of woodland in Australia, southern Africa, northeastern United States, Burma, and India. Knowing little of methods of sustained-yield forestry, they called in the Germans, who did. Thus in 1864 they established the Indian Forest Department, which for the first twenty-five years of its existence was serenely guided by three German inspectors general of forest-Dietrich Brandis, Wilhelm Schlich, and Bertold von Ribbentrop. For its effective functioning, the new department required a progressive curtailment of the previously untrammeled rights of use exercised by rural communities all over South Asia. An act was hurriedly drafted to establish the claims of the state to the forestland it immediately required, subject to the provision that existing rights not be abridged. This act was "infinitely milder and less stringent than that which is in force in most European countries." The search commenced for a more stringent and inclusive piece of legislation. In 1869, the Government of India circulated to the provinces a new draft act, which sought to strengthen the state's control over forest areas through the regulation and in some cases extinction of customary rights.

The new legislation was based on the assumption that all land not actually under cultivation belonged to the state. Of course, it was not easy to wish away the access to forests exercised in centuries past by peasants and other rural groups. The colonial state, however, argued that such use, however widespread and enduring, had been exercised only at the mercy of the monarch. Unless it had been expressly recorded in writing, customary use was deemed to be a "privilege," not a "right." And since the British government was the successor to Indian rulers, the ownership of forests and waste was now vested in it. "The right of conquest is the strongest of all rights," emphatically remarked one forest official. "It is a right against which there is no appeal."

There were, however, some notable dissenting voices within the colonial government. Sent the draft forest bill by the Government of India, the Madras Government in turn invited responses from various officers. The views of Narain Row, Deputy Collector of Nellore, are representative. The proposed legislation, he said, had no historical precedent, for "there were originally no Government forests in this country. Forests have always been of natural growth here; and so they have been enjoyed by the people."

Another Deputy Collector, Venkatachellum Puntulu, of Bellary, argued that the burden of the new legislation would fall most heavily on the poor. While large landlords would find it relatively easy to deny the state any claim over their forest property, unlettered peasants would not be able to prove rights of ownership, even though they traditionally used forests as common property.

Criticizing the detailed rules prohibiting the collection of different kinds of forest produce, Puntulu penetratingly remarked that "the provisions of this bill infringe the rights of poor people who live by daily labour (cutting wood, catching fish and eggs of birds) and whose feelings cannot be known to those whose opinions will be required on this bill and who cannot assert their claims, like [the] influential class, who can assert their claims in all ways open to them and spread agitation in the newspapers."

After several such responses came in, the Madras Board of Revenue told the Government of India that the claim of the state to uncultivated forests and wastes was virtually nonexistent:

There is scarcely a forest in the whole of the Presidency of Madras which is not within the limits of some village and there is not one in which so far as the Board can ascertain, the state asserted any rights of property unless royalties in teak, sandalwood, cardamoms and the like can be considered as such, until very recently. All of them, without exception are subject to tribal or communal rights which have existed from time immemorial and which are as difficult to define as they are necessary to the rural population.... [In Madras] the forests are, and always have been common property, no restriction except that of taxes, like the Moturpha [tax on tools] and Pulari [grazing tax] was ever imposed on the people till the Forest Department was created, and such taxes no more indicate that the forest belongs to the state than the collection of assessment shows that the private holdings in Malabar, Canara and the Ryotwari districts belong to it.

The Madras Government advanced three basic reasons for rejecting the bill drafted by the Government of India: First, because its principles, scope and purpose are inconsistent with the existing facts of forest property and its history. Second, because, even if the Bill were consistent with facts, its provisions are too arbitrary, setting the laws of property at open defiance, and leaving the determination of forest rights to a Department which, in this Presidency at all events, has always shown itself eager to destroy all forest rights but those of Government. Third, because a Forest Bill, which aims at the regulation of local usages ought to be framed.

The objections were disregarded, and in 1878, the new bill passed. The act divided the forests of the subcontinent into three broad classes. State or reserved forests were to be carefully chosen, in large and compact areas that could lend themselves to commercial exploitation.

The constitution of these reserves was to be preceded by a legal settlement that either extinguished customary rights of user, transferred them as "privileges" to be exercised elsewhere, or, in exceptional cases, allowed their limited exercise. In the second class, of "protected" forests, rights and privileges were recorded but not settled. However, all valuable tree species were to be declared as "reserved" by the state, while the Forest Department had the power to prohibit grazing and other ostensibly damaging practices.

The Forest Act also provided for a third class of forests-village forests. But as these lands had first to be constituted as reserved forests, the procedure aroused suspicion among the villagers, and this chapter remained a "dead letter."15 Meanwhile, the area of forests under strict state control steadily expanded. In 1878, there were 14,000 square miles of state forest. By 1890, this had increased to 76,000 square miles, three-fourths of which were reserved forests. Ten years later, there were 81,400 square miles of reserved forests and

8,300 of protected forests. Given increasing demand for wood products, the state sought to establish firmer control over forests, both by expanding the area taken over under the Forest Act and by converting protected forests to reserved forest. The Indian Forest Act of 1878 was a comprehensive piece of legislation that came to serve as a model for other British colonies.16 Within India, it allowed the state to expand the commercial exploitation of the forest while putting curbs on local use for subsistence. This denial of village forest rights provoked countrywide protest.

REBELLIONS AGAINST COLONIAL FORESTRY

The history of colonial rule is punctuated by major rebellions against colonial forestry-in Chotanagpur in 1893, in Bastar in 1910, in Gudem-Rampa in 1879-80 and again in 1922-23, in Midnapur in 1920, and in Adilabad in 1940. These rebellions sometimes extended over several hundred square miles of territory, involved thousands of villagers, and had to be put down by armed force. Even where discontent did not manifest itself in open rebellion, it was expressed through arson, noncompliance, and breaches of the forest law.

The participants in these protests were unlettered peasants and tribals, and we know far more of their deeds than their words.

Nonetheless, their voices do figure here and there in the archives of the state, sometimes mediated by the language of the officials reporting them. Thus in the 1880s, when the government of the Bombay Presidency was aggressively demarcating the rich teak forests of the Dang district, preparatory to their constitution as state reserves, a Bhil tribal chief sent in a petition stating that "we do not wish to let the Dang jungle [be] demarcated, for thereby we shall lose our rights and we and our poor rayat [cultivators] shall always be under the control of the Forest Department and the Department will always oppress us."

Around the same time, the colonial state was attempting to take over the deodar (cedar) forests of the upper Jamuna valley. These trees had suddenly become market-worthy, to service the then expanding railway network. But as a peasant bitterly observed, "the forests have belonged to us from time immemorial, our ancestors planted them and have protected them; now that they have become of value, government steps in and robs us of them."18 Or consider, finally, these remarks of an administrator in the Bastar district of central India, on the determination of his tribal subjects to continue practicing swidden cultivation in what was now "government" forests: "On the road from Tetam to Katekalyan I found general dissatisfaction at the restriction of penda [swidden] cultivation. I was unable to convince them of its evils [sic]. Podiyami Bandi Peda of Tumakpal has to get his son married and for this purpose he wants to cultivate penda in the prohibited area. I told him he should not do it. He replied plainly that he would cultivate it and go to jail as he had to get his son married."

In 1871, the Madras Government predicted that the new act, if passed into law, would "place in antagonism to Government every class whose support is desired and essential to the object in view, from the Zamindar [landlord] to the Hill Toda or Korombar." This was an astonishingly accurate prediction, for the Forest Department was unquestionably the most unpopular arm of the British Raj.

The story of the numerous popular movements against state forestry, so long neglected by historians, is now attracting an array of chroniclers. The critics were principally of two kinds. On the one side were scholars and politicians with a deep knowledge of rural conditions, and who sometimes formed part of popular movements themselves.

Their criticisms of state forestry thus drew richly upon the feelings and grievances of the people most affected by it. On the other side were the rare (but, for that reason, significant) dissidents within the colonial bureaucracy, who opposed the centralizing thrust of government forest policy. The first set were outsiders so far as the apparatus of rule was concerned, but insiders with respect to popular opinion and popular consciousness. The second set were by virtue of race and status outsiders to Indian society, but insiders with regard to the policy of the state and the functioning of government.

PRECOCIOUS PROPHETS

In 1878, the Poona Sarvajanik Sabha, a vastly respected nationalist organization in western India, bitterly opposed the new Forest Act. Despite its middle-class origins, the Sabha had consistenly fought for the rights of the cultivator, urging that the colonial government lessen its burden of taxation on the peasantry.21 Now, in the context of the Forest Act debate, it pointed out that state usurpation was grossly violative of customary rights over forests, for both "private grantees and village and tribal communities" had "cherished and maintained these rights with the same tenacity with which private property in land is maintained elsewhere." The Sarvajanik Sabha did not, however, merely oppose the proposed Forest Act for its excessive emphasis on state control; it offered a more constructive and creative alternative.

Thus the Sabha argued that the better maintenance of forest cover could more easily be brought about by taking the Indian villager into confidence of the Indian Government. If the villagers were rewarded and commended for conserving their patches of forestlands, or for making plantations on the same, instead of ejecting them from the forestland that they possess, or in which they are interested, emulation might be evoked between neighboring villages. Thus more effective conservation and development of forests in India might be secured, and when the villagers have their own patches of forest to attend to, government forests might not be molested. Thus the interests of the villagers as well as the government can be secured without causing any unnecessary irritation in the minds of the masses of the Indian population.

The Sabha was advocating a far more democratic structure of forest management than that envisaged by the colonial government. Indeed, it was proposing the institution of a Vrikshamitra (Friends of the Trees) Award, one hundred and ten years before the Indian Government's Ministry of Environment and Forests conceived and named such a scheme, for rewarding individuals and communities who had successfully protected or replenished forest areas. Three years after the 1878 act was passed, the impact of state forestry on rural communities was foregrounded by the social reformer Jotirau Phule. Phule himself was a gardener by caste, and in general exceptionally alert to the problems of the agricultural classes. The following is his description of the impact of the Forest Department on the livelihood of farmers and pastoralists in the Deccan countryside: In the olden days small landholders who could not subsist on cultivation alone used to eat wild fruits like figs and jamun and sell the leaves and flowers of the flame of the forest and the mahua tree [both common trees of the Indian forest].

They could also depend on the village ground to maintain one or two cows and two or four goats, thereby living happily in their own ancestral villages. However, the cunning European employees of our motherly government have used their foreign brains to erect a great superstructure called the forest department. With all the hills and undulating areas as also the fallow lands and grazing grounds brought under the control of the forest department, the livestock of the poor farmers do not even have place to breathe anywhere on the surface of the earth.

DEPENDENCE OF THE AGRICULTURIST ON THE PRODUCE OF FORESTS

These remarks drew attention to the dependence of the agriculturist on the produce of forests and other common lands. This dependence was even more acute in the tribal regions of middle India, where communities of hunter-gatherers, swidden agriculturists, and charcoal iron makers were likewise at the receiving end of the new forest laws. These peoples found an eloquent spokesman in Verrier Elwin (1902-1964), a brilliant Oxford scholar and renegade priest who became the foremost interpreter of adivasi (tribal) culture in India. Elwin was a pioneer of ecological anthropology, whose many works vividly showcased the intimate relationship between the forest world and the life of the adivasi. All tribals, he argued, had a deep knowledge of wild plants and animals; some could even read the great volume of Nature like an "open book." Swidden agriculturists, for whom forest and farm shaded imperceptibly into each other, had an especial bond with the natural world. They liked to think of themselves as children of Dharti Mata, Mother Earth, fed and loved by her.

Elwin's ethnographies are peppered with references to the adivasi's love for the forest. Tragically, the forest and game laws introduced by the British had made them interlopers in their own land. He quotes a member of the tribal group Gond, whose idea of heaven was "miles and miles of forest without any forest guards." As the anthropologist himself wrote in 1941:

The reservation of forests was a very serious blow to the tribesman. He was forbidden to practice his traditional methods of cultivation. He was ordered to remain in one village and not to wander from place to place. When he had cattle he was kept in a state of continual anxiety for fear they should stray over the boundary and render him liable to heavy fines. If he was a Forest Villager he became liable at any moment to be called to work for the Forest Department. If he lived elsewhere he was forced to obtain a license for almost every kind of forest produce. At every turn the Forest Laws cut across his life, limiting, frustrating, destroying his self confidence.

During the year 1933-4 there were 27,000 forest offences registered in the Central Provinces and Berar and probably ten times as many unwhipped of justice. It is obvious that so great a number of offences would not occur unless the forest regulations ran counter to the fundamental needs of the tribesmen. A Forest Officer once said to me: "Our laws are of such a kind that every villager breaks one forest law every day of his life."

Elwin's writings were addressed equally to the colonial state and to the Congress nationalists, who in the 1940s were very much a government-in-waiting. The Congress, however, had not been especially sensitive to the rights of the tribals. But as Elwin reminded them, "the aboriginals are the real swadeshi [indigenous] products of India, in whose presence everything is foreign. They are the ancient people with moral claims and rights thousands of years old. They were here first: they should come first in our regard." He was deeply distressed when a Congress report on tribals followed the British authorities in asking for a ban on shifting cultivation.

Now Elwin's work had shown that, contrary to modernist prejudice, swidden as practiced by the Baiga, the Juang, and other tribes was an ecologically viable system of cultivation. When the nationalists recommended the ban, he wrote angrily that "the forests belong to the aboriginal. I should have thought that anyone who was a Nationalist would at least advocate swaraj [freedom] for the aboriginal!"

The significance of the forest in tribal life is a running theme in Elwin's work. Noting that a majority of tribal rebellions had centered around land and forests, he pleaded for the greater involvement of tribals in forest management in free India. Even if adivasis had no longer any legal rights of ownership, they had considerable moral rights. And as tribals were as much part of the national treasure as forests themselves, there should be an amicable adjustment between forest management and tribal needs. Even where commercial forest operations became necessary, he said, these should be undertaken by tribal cooperatives and not by powerful private contractors. After independence Verrier Elwin became the first foreigner to be granted citizenship of free India. In 1954, he was appointed Adviser on Tribal Affairs to the Government of India (with special reference to the North-east Frontier Agency). He was also to serve on more than one high-level, all-India committee on tribal policy. From his first official appointment until his premature death in 1964, Elwin repeatedly urged a reconsideration of forest policy, such that it might, at last, come to more properly serve tribal needs. In this he had little success for forest management became, if anything, more commercially oriented in independent India.32 Towards the end of his life, the anthropologist wrote with some bitterness of how the victims of government policy were being unfairly blamed for the destruction of forests:

There is constant propaganda that the tribal people are destroying the forest. When this was put to some of the villagers, they countered the complaint by asking how they could destroy the forest. They owned no trucks; they hardly had even a bullock-cart; the utmost that they could carry away was a headload of produce for sale to maintain their families and that too against a license. The utmost that they wanted was wood to keep them warm in the winter months, to reconstruct or repair their huts and carry on their little cottage industries. Their fuel-needs for cooking, they said, were not much, for they had not much to cook. Having explained their own position they invariably turned to the amount of [forest] destruction that was taking place all around them. They asked how the exzamindars [landlords], in violation of their agreements and the forest rules and laws, devastated vast tracts of forest land right in front of officials. They also related how the contractors stray outside the contracted coupes, carry loads in trucks in excess of their authorised capacity and otherwise exploit both the forests and the tribal people.

ARGUMENTS IN FAVOR OF PRESERVATION AND DEVELOPMENT OF FORESTS

There is a feeling among the tribals that all the arguments in favor of preservation and development of forests are intended to refuse them their demands. They argue that when it is a question of industry, township, development work or projects of rehabilitation, all these plausible arguments are forgotten and vast tracts are placed at the disposal of outsiders who mercilessly destroy the forest wealth with or without necessity.

From a great Englishman who devoted his life to the service of the Indian poor, we move on to a great Englishwoman who did likewise-Madeleine Slade, the daughter of an admiral who came from England to join Gandhi in his Sabarmati Ashram in 1926. Gandhi adopted her as his own daughter and gave her the name Mira Behn. She played a prominent role in the anti-colonial struggle and was jailed several times. In 1945, Mira Behn set up a Kisan (peasant) Ashram near the holy town of Hardwar, and two years later moved up the Ganges beyond Rishikesh, where the river descends into the plains. In

1952, she shifted her base again, to the Bhilangna valley in the interior Himalaya. Here she stayed still 1959 when ill health and possibly dissatisfaction with the policies of independent India made her migrate to Austria.

The peasants of the central Himalaya are, of course, as dependent on forest produce as the tribals of the Indian heartland with whom Elwin long worked. Here, one unfortunate consequence of state forest management was the gradual replacement of banj oak (quercus incana), a tree much prized by villagers as a source of fuel, fodder, and leaf manure, by chil (or chir) pine (pinus roxburghii), a species more valued commercially as a source of timber and resin. This transition had serious ecological implications, for the thick undergrowth characteristic of banj forests absorbed a high proportion of the rainwaters of the fierce Himalayan monsoon. This water then slowly percolated downhill. Below the oak forests were thus found "beautiful sweet and cool springs," the main source of drinking water for the hill villagers. By contrast, the floor of pine forests was covered thinly by needles, and had much less absorptive capacity. In hillsides dominated by chil, the rain rushed down the slopes, carrying away soil, debris, and rock, contributing thereby to floods.

Why were the banj forests disappearing in the Himalaya? Mira Behn's own explanation revealed a sharp awareness of the sociology of forest management in the hills. "It is not merely that the Forest Department spreads the Chil pine," she said, "but largely because the Department does not seriously organize and control the lopping of the Banj trees for cattle fodder, and... is glad enough from the financial point of view to see the Banj dying out and the chil pine taking its place. When the Banj trees grow weak and scraggy from overlopping, the chil pine gets a footing in the forest, and once it grows up and starts casting its pine needles on the ground, all other trees die out."

Mira Behn continued: "It is no good putting all the blame on the villagers.... The villagers themselves realize fully the immense importance of these Banj forests, without which their cattle would starve to death, the springs would dry up, and flood waters from the upper mountain slopes would devastate their precious terraced fields in the valleys. Indeed all these misfortunes are already making their appearance on a wide scale. Yet each individual villager cannot resist lopping the Banj trees in the unprotected Government forests. 'If I do not lop the trees someone else will, so why not lop them, and lop them as much as possible before the next comer.'"

Although Mira Behn does not explicitly make the point, it seemed that this shortsighted behavior of the hill peasant was related to the loss of community control, such that individual peasants no longer had a long-term stake in the maintenance of forest cover. This was a tendency aggravated by the commercial orientation of the Forest Department. Could anything be done to restore banj to its rightful place, and thus revive Himalayan economy and ecology? Mira Behn writes: The problem is not without solution, for if trees are lopped methodically, they can still give a large quantity of fodder, and yet not become weak and scraggy. At the same time, if the intruding Chil pines are pushed back to their correct altitude (*i.e.* between 3,000 and 5,000 feet), and the Banj forests are resuscitated, the burden on the present trees will, year by year, decrease, and precious fodder for the cattle will actually become more plentiful.

But all this means winning the trust and co-operation of the villagers, for the Forest Department, by itself, cannot save the situation. Nor can it easily win the villagers' trust, because the relations between the Department and the peasantry are very strained, practically amounting to open warfare in Chil pine areas. Therefore, in order to awaken confidence in the people, some non-official influence is necessary. With the aid of local constructive workers, it should become possible to organize village committees and village guards to function along with the Forest Department field staff which should be increased, and also given special training in a new outlook towards the peasantry.

In this way it should be feasible to carry out a wellbalanced long term project for controlled lopping and gradual return of the Banj forests to their rightful place, by systematic removal of Chil pines above 5,500 feet altitude to be followed by protection of the young Banj growth. The Banj forests are the very centres of nature's economic cycle on the southern slopes of the Himalayas. To destroy them is to cut out the heart and thus bring death to the whole structure.

Mira Behn sent reports of her findings, with photographs, to Prime Minister Jawaharlal Nehru. He passed them on to the concerned officials, but nothing seems to have come of it; it appears that the Indian Forest Department of the 1950s would not change its ways.

A DEMOCRATIZING FORESTER

To those who know something of the people behind them, they are also perfectly in character. Phule, Elwin, and Mira Behn, as well as the leaders of the Poona Sarvajanik Sabha, had a deep knowledge of agrarian life. Alert to the inequities in access to natural resources brought about by the new laws, they would vigorously polemicize on behalf of the victims of state forest management. Dietrich Brandis was a prophet of community forestry who came from the unlikeliest of backgrounds. He was a forest officer; in fact, no less than the first inspector general of forests (IGF) in India. In nineteen years as IGF (1864-1883), he laid the foundations of state forestry in India.

A man of great energy, he toured widely in the subcontinent, writing authoritative reports on the direction forest management should take in the different provinces of British India.36 In the realm of silviculture, he formulated the systems of valuation and forest working still widely in use. As a former university don himself (he came to the service of the Raj from the University of Bonn), Brandis started a college for training subordinate staff, arranged for higher officials to be trained on the continent, and helped set up the Forest Research Institute in Dehradun.

The scientific and administrative aspects of Brandis's legacy are not our focus here, but rather his sociology of forest management, his understanding of the social and political contexts within which state forestry had to operate in India.37 Here Brandis's views must be immediately distinguished from almost all other forest officials, Indian or European, before or since. These officials counterpose "scientific" forestry under state auspices to the customary use of forests by rural communities, which they have always held to be erratic, unsystematic, wasteful, and shortsighted. It is thus that the forest officials justify their territorial control of over one-fifth of India's land mass, claiming that they alone possess the technical skills and administrative competence to manage woodland.

To be sure, Dietrich Brandis shared this creedal faith in the scientific status of sustained-yield forestry. He also believed that the state had a central role to play in forest management. But what he certainly did not share was his colleagues' skepticism of the knowledge base of rural communities. For example, Brandis wrote appreciatively of the widespread network of sacred groves in the subcontinent. These he termed, on different occasions, the "traditional system of forest preservation" and examples of "indigenous Indian forestry." In his tours he found sacred woodlands "most carefully protected" in many districts-from the Devara Kadus of Coorg in the south to the deodar temple groves in the Himalaya.

At the other end of the social spectrum, Brandis also wrote appreciatively of forest reserves managed by Indian chiefs. He was particularly impressed by the Rajput princes of Rajasthan, whose hunting preserves provided game for the nobility as well as a permanent supply of fodder and small timber for the peasantry. The British stereotype of the Indian Maharaja was of a feckless and dissolute ruler, but as Brandis pointed out, in strenuously preserving brushwood in an arid climate the Rajputs had "set a good example, which the forest officers of the British government would do well to emulate."

In Brandis's larger vision for Indian forestry, a network of state reserves would run parallel to a network of village forests. The Forest Department would take over commercially valuable and strategically important forests, while simultaneously encouraging peasants to collectively manage areas left out of these reserves. Through a series of reports and memoranda written over a decade, the IGF tried to persuade the colonial government that a strong system of village forests was vital to the longterm success of state forestry itself. The first such report was written in 1868, and pertained to the southern province of Mysore. This was a closely argued document suggesting the creation of village forests throughout Mysore, managed on a rotational cropping system, with freshly cut areas closed to fire and grazing. Ideally, each hamlet would have its own forest, but in many cases it might become necessary to constitute a block to be used by a group of villages. Such forests would provide the following items free of cost-firewood for home consumption and for sale by "poor people with headloads"; wood for agricultural implements and the making and repairing of carts; wood, bamboo, and grass for thatching, flooring, and fencing; leaves and branches for manure; and grazing except in areas closed for reproduction. On the payment of a small fee, wood would be made available for houses and for use by artisans.

In Brandis's scheme, these forests would be put under a parallel administrative system, with a village forester for each unit, a forest ranger for all the village forests in each taluk (county), and a head forest ranger for the district as a whole, this man reporting to an assistant conservator of forests. He anticipated that the system would be self-supporting, with any surplus used for local improvements. In this manner, peasants would come to feel an interest "in the maintenance and improvement of their forests." Brandis also hoped that Forest Department control over village forests would give way in due course, with the "leading men" of each village assuming responsibility for management.

Forwarding his report to the Government of India, Brandis noted significantly that it was "the first of a series of measures" which he proposed "to suggest in various Provinces for the better utilization and for the improvement of the extensive wastelands which will not be included in the State Forests": that is, as a prelude to recommending a countrywide system of community forests. Unhappily, the British officials of the Raj lacked Brandis's understanding of the biomass economy of rural India, the vital dependence of agrarian life on the produce of the forests. They also lacked his faith in local knowledge and local initiative.

The opposition to Brandis claimed that his scheme would lead to a loss of state revenues while undermining the powers of district officials. Also invoked was an early version of the "Tragedy of the Commons" argument. For one official, "the village communities of Mysore, without cohesion and often split up into factions by caste, could not be entrusted with the powers, or competent to perform the functions assigned to them in [Dr Brandis's] scheme." Another commented that the scheme would fail "as each man, when the least removed from supervision, would cut whatever he might require for himself without any regard to the interests of his neighbours." The Government of India's final, negative verdict rested on a classic piece of colonial stereotyping. "The prejudices and rivalries of Natives," it said, "might be excited if men of different classes and castes shared in the same forests."

Brandis did not lightly accept this judgment. In a defiant note, he reviewed the case afresh, and made another forceful plea in favor of village forests. He drew pointed attention to the flourishing system of community forests on the continent, where scientific foresters exercised technical supervision over woodland managed for the exclusive benefit of villages and small towns. In

Europe, wrote Brandis, "Such Communal Forests are a source of wealth to many towns and villages in Italy, France and Germany; property of this nature maintains a healthy spirit of independence among agricultural communities; it enables them to build roads, churches, school-houses, and to do much for promoting the welfare of the inhabitants; the advantages of encouraging the growth, and insisting on the good management of landed communal property, are manifold, and would be found as important in many parts of India as they have been found in Europe."

Following his failure in Mysore, Brandis resurrected his proposals in the debate leading up to the 1878 Forest Act. Here he urged the administration "to demarcate as state forests as large and compact areas of valuable forests as can be obtained free of forest rights of persons," while leaving the residual area, smaller in extent but more conveniently located for their supply, under the control of village communities. He hoped ultimately for the creation of three great classes of forest property, based on the European experience: state forests, forests of villages and other communities, and private forests. State ownership had to be restricted, argued Brandis, on account of the "small number of experienced and really useful officers" in the colonial forestry service and out of deference to the wishes of the local population. For "the trouble of effecting the forest rights and privileges on limited well-defined areas is temporary and will soon pass away, whereas the annoyance to the inhabitants by the maintenance of restrictions over the whole area of large forest tracts will be permanent, and will increase with the growth of population."

Here was an uncanny anticipation of the widespread popular opposition that has been such a marked feature of the subsequent history of Indian forestry. But Brandis was overruled by more powerful civil servants within the colonial bureucracy, and the 1878 Forest Act was based firmly on the principle of state monopoly. But the German forester was a remarkably persistent man. As he remarked shortly after relinquishing the post of inspector general, systematic forestry in India "was like a plant of foreign origin, and the aim must be to naturalize it."

On the social side, this process of indigenization could be accomplished by encouraging native chiefs, large proprietors, and especially village communities to develop and protect forests for their own use. In the last instance the initiative lay with the government, which, insisted Brandis, stood to gain enormously from a successful system of communal forests. "Not only will these forests yield a permanent supply of wood and fodder to the people without any material expense to the State," he wrote, "but if well managed, they will contribute much towards the healthy development of municipial institutions and of local self-government."

In 1897, well into his retirement in Germany, Brandis returned to the subject of community forests. Long after he had severed all formal contacts with British India, Brandis continued to be deeply concerned that Indian forestry should cease to have "the character of an exotic plant, or a foreign artificially fostered institution." This concern was consistent with his larger democratic vision for forestry in the subcontinent. Thus he suggested that "native" forest officers, as they distinguished themselves, be sent to study the forestry system operating in Germany. Notably, Brandis had in mind their social as well as silvicultural education. As he concluded his essay, Indian foresters, if sent to Germany, "will find that the villages, which own well-managed communal forests, are prosperous, although now and then they complain of the restrictions that a good system of management unavoidably imposes. What Indian forest officers will learn in this respect in Germany will be really useful to them in India."

Perhaps by now Brandis despaired of British officials in India taking seriously his proposals for the constitution of village forests. Hence this indirect approach, wherein Indian forest officers trained on the continent might be able to better see the benefits of community forests. In the event, Indian officials (whether trained in Germany or not) have been, for the most part, hostile to any suggestion that local communities could be encouraged to manage forest areas for their own use. It is, indeed, this territorial monopoly and indifference to the demands of rural communities that have made the Forest Department the object of such relentless criticism in recent years.

The Indian Forest Department has been the subject of sharp attack for its authoritarian style of functioning; and yet, in an interesting paradox, the founder of the department had himself anticipated that a narrow reliance on state control and punitive methods of management would lead to popular disaffection. While terms such as "social forestry," "community forestry," and "joint forest management" have only now come into currency, the principles they embody would have been readily recognized, and indeed warmly commended, by the first head of the Forest Department in India.

THE HIMALAYAN FOREST

For all their insight, knowledge, and passion, these precocious advocates of community forestry did not have much impact on state policy. Control and commercialization remained the dominant motifs of state forest policy. The chapter on village forests in the 1878 act remained a dead letter. Government forest policy, in the colonial as well as postcolonial periods, continued to seriously ignore village needs, demands, and interests. The principle of state monopoly has remained paramount, with one very partial exception. The Kumaun and Garhwal hills of present-day Uttar Pradesh contain the best stands of softwood in the subcontinent. These coniferous species have been highly prized since the early days of colonial forest management. Between 1869 and 1885, for example, some 6.5 million railway sleepers made from deodar (cedrus deodarus, the Indian cedar) were exported from the valley of the Yamuna, in the princely state of Tehri Garhwal. Adjoining Tehri Garhwal to the east was the British-administered Kumaun division, with its rich stands of chir pine. Here forestry operations concentrated simultaneouly on expanding the area under chir (at the expense of oak) and exploiting the tree both for timber and for resin. Between 1910 and 1920, for example, the number of trees tapped for resin increased from 260,000 to 2,135,000. The pine trees of the Central Himalaya were the only source, within the British Empire, of oleo-resin, an extract with a wide range of commercial and industrial applications. Likewise, the timber of deodar and chir, as well as fir and spruce, constituted a strategically valuable resource for the colonial state, exploited with profit to service the military campaigns of the two world wars.

In the Himalaya, as elsewhere, commercial forestry under state auspices was made possible only through a denial of customary rights of ownership and use. In these hills, forests and grassland were a crucial resource for the agro-pastoral production system.

In fact, the fragmentary evidence available to historians does suggest the existence of a fairly widespread system of common property resource management-with grass reserves walled in and well looked after, oak forests managed by the village community, and sacred groves lovingly protected. Not surprisingly, the government's attempts to seize vast areas under local control and reconstitute them as "reserved forests" evoked opposition. In the early years of state management, a petition from a discontented hillman evocatively recalled a golden age when the villagers had full control over their forest habitat: In days gone by every necessities of life were in abundance to villagers than to others [and] there were no such government laws and regulations prohibiting the free use of unsurveyed land and forest by them as they have now. The time itself has now become very hard and it has been made still harder by the imposition of different laws, regulations, and taxes on them and by increasing the land revenue. Now the village life has been shadowed by all the miseries and inconveniences of the present day laws and regulations.

They are not allowed to fell down a tree to get fuels from it for their daily use and they cannot cut leaves of trees beyond certain portion of them for fodder to their animals. But the touring officials still view the present situation with an eye of the past and press them to supply good grass for themselves and their [retinue] without thinking of making any payment for these things to them who after spending their time, money and labour, can hardly procure them for their own use. In short all the privileges of village life, as they were twenty years ago, are nowhere to be found now, still the officials hanker after the system of yore when there were everything in abundance and within the reach of villagers.

When such protests went unheeded, the sentiments underlying them were to manifest themselves in sustained and organized resistance on the part of the Himalayan peasantry. In fact, this region probably witnessed more and more serious social conflict than any other forest region of India.

PEASANT MOVEMENTS AGAINST STATE FOREST POLICIES

There were major peasant movements against state forest policies in 1904, 1906, 1916, 1921, 1930, and 1942. These recurrent conflicts, remarked one sensitive official, were a consequence of "the struggle for existence between the villagers and the Forest Department; the former to live, the latter to show a surplus and what the department looks on as efficient forest management." The most significant forest movement in Kumaun and Garhwal took place in 1921. This took the shape of labour strikes, which crippled the administration, and the widespread burning of pine forests. A total of 395 recorded fires burnt an estimated 246,000 acres of forest. Hundreds of thousands of resin channels were destroyed.

Constituting a direct challenge to the state to relax its control over forest areas, these protests enjoyed enormous popular support, which made it virtually impossible for the administration to detect the people responsible for the fires. The fires were generally directed at areas where the state was at its most vulnerable, for example, compact blocks of chir forest worked for timber and/or resin. Significantly, there is no evidence that the large areas of broad-leaved forest, also controlled by the state, were at all affected. Thus arson was not random but carefully discriminating-it spared those species more useful to the village economy.

In the vanguard of the 1921 movement were soldiers who had fought for the British in the First World War. Kumaun and Garhwal had long supplied hardy and exceptionally brave soldiers for the British Army-indeed, three of the five Victoria Crosses awarded to Indians between 1914 and 1918 went to this region.

These former men in uniform saw the forest regulations as a bitter betrayal of their interests by the white overlord for whom they had so recently risked their lives. Their protests alarmed the colonial state, for apart from being a reservoir of able-bodied men whom it hoped to continue to recruit for the wars it had to fight, the Kumaun hills bordered both Nepal and Tibet-regions not under direct British suzerainty but in which it had strong trading and political interests. In the wake of the popular protests, a magisterial critique of government forest policy was published by Govind Ballabh Pant. Pant was a rising lawyer from a peasant household in Almora who went on to become one of the foremost of Indian nationalists, after independence taking office successively as chief minister of Uttar Pradesh and home minister of the Government of India.

His 1922 booklet The Forest Problem in Kumaun described the "burial of the immemorial and indefeasible rights of the people of Kumaun," buried, that is, "between the property-grabbing zeal of the revenue officers and the exhortations of experts of the forest department." As he put it, with legal precision, "the policy of the Forest Department can be summed up in two words, namely, encroachment and exploitation." Several decades of a single-minded commercial forestry had led to a manifest deterioration of the agrarian economy: "Symptoms of decay are unmistakably visible in many a village: buildings are tottering, houses are deserted, population has dwindled and assessed land has gone out of cultivation since the policy of [forest] reservation was initiated.... Cattle have become weakened and emaciated and dairy produce is growing scarce every day: while in former times one could get any amount of milk and other varieties for the mere asking, now occasions are not rare when one cannot obtain it in the villages, for any price for the simple reason that it is not produced there at all."

Pant's analysis was rooted in a deep knowledge of the local context. He took it upon himself to combat the charge, commonly levied against the hill peasant, "of reckless devastation," a charge "sedulously propagated by prejudiced or ignorant persons." As he wrote, The spacious wooded areas extending over the mountain ranges and hill sides bear testimony to the care bestowed by the successive generations of the Kumaonies.

All of them are not of spontaneous growth and specially the finer varieties bespeak his labour and instinct for the plantation and preservation of the forest. A natural system of conservancy was in vogue, almost every hill top is dedicated to some local deity and the trees on or about the spot are regarded with great respect so that nobody dare touch them.

There is also a general impression among the people that everyone cutting a tree should plant another in its place.... Grass and fodder reserves are maintained, and even nap [cultivable] lands are covered with trees, wherever, though in few cases, such land could be spared from the paramount demand of cultivation. Special care is also taken by the villagers to plant and preserve trees on the edges of their fields. From this analysis, the solution logically offered was to give back to the peasants the woodland that they traditionally regarded as being within their village boundaries.

"If the village areas are restored to the villagers, the causes of conflict and antagonism between the forest policy and the villagers will disappear, and a harmony and identity of interests will take the place of the distrust, and the villager will begin to protect the forests even if such protection involves some sacrifice or physical discomfort." Pant envisaged that these areas would be under the control of the village panchayats, or councils, under whose direction the "natural system of conservancy" would once again come to the fore. As he shrewdly observed, "some restrictions will be there, but these will proceed from within, and will not be imposed from without."

Clearly, Pant drew upon and systematized the knowledge, perceptions, and analysis of the peasant folk of the Kumaun Himalaya. Thus, after the popular protests in 1921, the Government of the United Provinces set up a Kumaun Forest Grievances Committee. This committee toured the hills, examining some five thousand witnesses in all.

Peasant activists submitted dozens of petitions to the committee, on behalf of individual villages. These identified blocks of forest near every village, where peasants would have exclusive rights of fuel and fodder collection, timber for building, wood for ploughs, bamboos for basket making, etc. It was being proposed that villagers should have full rights over these forests, which they would manage through their own panchayat. Based on the evidence it collected, the committee finally concluded that "any attempt to strictly enforce these [forest] rules would lead to riots and bloodshed."

It thus divided the existing reserved forests into two categories-Class I, which were to be managed not by the forest officials but by the civil administration (in theory more sympathetic to rural needs), and Class II, constituting the commercially valuable wooded areas, which were to remain with the Forest Department. It also recommended that the government consider the constitution of village forests as per the demands of the people of Garhwal and Kumaun.

Bureaucracies move at their own pace, and only in 1930 the rules were passed allowing for the formation of van panchayats, or village forests, in the hill districts of the United Provinces. These allowed for a forest patch to be handed over to a village if it lay within its settlement boundaries, and if more than one-third of its residents had applied for permission to the deputy commissioner (DC). Once the DC gave the go-ahead, then the villagers elected, by voice vote, a council (panch) of five to nine members. This council in turn elected a head (sarpanch) among themselves. The van panchayat was empowered to close the forest for grazing, regulate cutting of branches and collection of fuel, and organize the distribution of forest produce.

PERMISSION OF THE FOREST DEPARTMENT

It could appoint a watchman, whose salary would be paid by villagers' contributions to the panchayat. The panchayat could levy fines, although if the offender did not pay it had then to go to the civil courts for redressal.

The felling of trees, however, required the permission of the Forest Department. The department also claimed 40 percent of the revenue from any commercial exploitation. Of the rest, 20 percent would go to the zilla parishad (district council), with the balance 40 percent kept with the DC on behalf of the van panchayat, which with that official's written consent could use the funds for roads, schools, and other local improvements.

There are now in excess of 4,000 van panchayats in Kumaun and Garhwal, covering an area of just less than half a million hectares. An official report of 1960 remarked that many of these village forest councils had done "exemplary work in connection with forest protection and development." A more recent survey has concluded that the panchayat forests are often in a better condition than the reserved forests. Of twenty-one panchayats surveyed in three districts, the forest stock in thirteen of these were in good condition, in four in medium condition, in three in poor condition. The researcher concluded that van panchayats have, by and large, maintained oak forests very well, especially in contrast to the dismal condition of the reserves (except for those reserves distant from habitations). The position in respect of chir forests is not so clear, but these seem to have done about as badly under van panchayat control as in the reserves. Various studies suggest that, overall, panchayat forests seem to be in as good or better condition than the reserves.

The van panchayat system constitutes the only network of village forests mandated by law in all of India.61 The concession was made by a colonial state worried of losing control in a sensitive and strategically important border region, and it was not to be replicated elsewhere. After independence, the van panchayat regime was not extended to the adjoining region of Tehri Garhwal, where Mira Behn worked in the 1950s, and where it might have very well contributed to preserving and enhancing the oak forests. Within Kumaun, too, there is considerable resentment over the curbs placed on the autonomous functioning of van panchayats. Though technically under the control of the villagers, the Forest Department can veto schemes for improvement, while of the revenue generated, 40 percent is swallowed by the state exchequer. Forty percent of the rest is by law granted to the village, but this money too first finds its way into a "consolidated fund" controlled by the DC, to which individual panchayats have then to apply.

There are signs of an emerging movement to do away with these constricting rules, to make the management of the panchayats come fully under the control of the villagers. A chronicler of this discontent, himself quite aware of the long history of forest-related protests in Kumaun and Garhwal, writes that "those who know the history of forest struggles say that... the van panchayat movement will be the biggest such movement in the hills."

TWO CHEERS FOR JOINT FOREST MANAGEMENT

From the inception of state forestry in India, perceptive critics have argued for a democratization of resource control, for a correction of the commercial bias promoted by successive governments, and for a proper participation in management and decision making by local user groups. Arguments first offered in the 1870s, and reiterated in subsequent decades, were revived, or reinvented, in the 1970s by the now-famous Chipko movement. It is no accident that Chipko originated in Garhwal and Kumaun, the part of India that has seen some of the most intensive conflicts between the state and the peasantry over forest resources.

The 1970s were marked by a series of forest movements in different parts of India. These took place in the Himalaya, in the Western Ghats, and, above all, in the vast tribal belt extending across the heart of peninsular India. In the Chotanagpur plateau, forest protests formed an integral part of the larger movement for a separate tribal homeland of Jharkhand, carved out from the huge, unwieldy, and predominantly non-tribal state of Bihar. In one much celebrated case, tribals demolished a plantation of teak, a highly prized furniture wood, that was coming up on land previously under the sal tree (Shorea robusta), a species of far greater benefit to the local economy.

Their slogan, "Sal means Jharkhand, sagwan (teak) means Bihar" was a one-sentence critique of the narrow commercial ends of state forestry. Since the 1970s, there has been an ongoing, nationwide debate on forest policy in India, a debate fuelled by the continuing social tension in forest areas and the evidence of massive deforestation provided by satellite imagery.

This debate has passed through three distinct if chronologically overlapping phases. The first phase might be designated the "politics of blame." The activists speaking on behalf of disadvantaged groups have held the forest officials responsible for environmental degradation and popular discontent. The officials, in turn, have insisted that growing human and cattle populations are the prime reason why fully half of the 23 percent of India legally designated as "forest" was without tree cover.

The forestry debate of the 1970s and the 1980s drew, at times, on the heritage of earlier movements and critiques. The peasants of Garhwal and Kumaun, as this writer found out while doing field work there in 1982-83, were acutely conscious of how Chipko itself drew on a long and honorable history of peasant resistance to state forestry.

Tribal activists in Madhya Pradesh and Bihar, meanwhile, were not unfamiliar with the work and message of Verrier Elwin. And in the villages of the Deccan, social workers liked to offer the same quote of Jotirau Phule's reproduced earlier in the paper, as proof that in the agro-pastoral system of that region, proper access to forests and pasture was vital to survival, and that it was the "great superstructure" of the Forest Department that continued to deny herders and farmers this access.

However, perhaps the most direct connection between the past and the present of forest management was effected in the summer of 1982, when the Government of India circulated a new draft forest act. Activists and academics joined hands to demonstrate how the proposed legislation was solidly based upon and, indeed, took further forward the centralizing thrust and punitive orientation of the notorious Indian Forest Act of 1878. After a countrywide campaign, the draft bill was finally dropped by the state.

As tempers cooled and polemic exhausted itself, a second phase, the "politics of negotiation," originated. In villages and state capitals, forest officers and their critics found themselves at the same table, talking and beginning to appreciate, if not fully understand, the other's point of view. Concessions were made by each side, protests suspended by one, and leases of forest produce to industry cancelled by the other. One product of the growing dialogue between activists and bureaucrats was the approval, by the Indian Parliament in 1988, of a new National Forest Policy. Where the ruling Forest Policy of 1952 had stressed state control and industrial exploitation, the new document instead emphasized the imperatives of ecological stability and peoples' needs.

JOINT FOREST MANAGEMENT

Then, slowly and hesitatingly, commenced the third phase, "the politics of collaboration." In the state of West Bengal, for example, the Forest Department initiated remedial action on its own, abandoning its traditional custodial approach by inviting peasants to cooperate with it. Thousands of village forest protection committees were constituted, each of which pledged to protect nearby forests in collaboration with the state. Thus previously authoritarian government officials joined with previously suspicious villagers to succesfully regenerate the degraded sal forests of southwestern Bengal.

The success of "Joint Forest Management" or JFM in West Bengal has encouraged scholars, activists, and sympathetic civil servants to demand its replication in other parts of India. Outside its original home, however, the progress of JFM has been slow. Administrative styles and cultures of governance vary widely among the states and regions of India. So do individual orientations, with some forest officials still loathe to relinquish control, while others have been inspired to start village protection committees on their own.

A mapping of the forestry debate in contemporary India would therefore show significant regional variations. Some states are still stuck in the "politics of blame"; others have moved tentatively to the "politics of negotiation." West Bengal and parts of Andhra Pradesh, Madhya Pradesh, and Himachal Pradesh have instituted the "politics of collaboration" through the creation of JFM regimes. In this last scenario there is abundant scope for improvement. As analysts have shown, the JFM model now promoted by the Government of India reflects and sometimes reinforces inequities within rural society.

Gender and caste are two axes of discrimination, with women and low-caste members of the village community not having adequate representation or voice in the decision-making process (this is also true, to a great extent, of the van panchayats in Kumaun.) Likewise, pastoral groups and artisans, who have legitimate claims on forest resources, are sometimes given short shrift. Moreover, the forest officials still claim a monopoly of "scientific expertise," refusing to entertain villagers' own ideas on species choice, spacing, or harvesting techniques.

One serious problem with the JFM model, as currently promoted by the state and donor agencies, is that it allows the constitution of village forest committees only on forestland with less than 40 percent crown cover. This is a deeply constricting rule, which reserves to the state, and the state alone, exclusive rights over the bestclothed lands of India. Thus forests situated close to hamlets cannot come under JFM regimes if they have more than 40 percent tree cover. Again, the regulations, strictly interpreted, would mean that if local communities were to effectively protect and replenish degraded lands, such that the crown cover was to come to exceed that magic figure of 40 percent, the state could step in and remove the area from JFM-which would be a bizarre outcome indeed.

Nor have changes in policy and orientation been accompanied by concomitant changes in legislation. Thus, the present regime is not flexible enough to allow for spontaneous community-initiated forest regimes to exist along with more orthodox JFM regimes. In some parts of India, the Forest Department is casting a covetous eye on areas well protected by village communities.

Thus in the Uttar Pradesh hills, the old established panchayat forests, managed by villagers, are sought to be brought under the JFM system only so that bureaucrats would have a greater say in their management. A new, carefully thought out Indian Forest Act is called for, which allows both for areas to be managed under state-village partnerships as well as by self-generated, autonomous community regimes.

One can thus envision a fourth (and possibly final) phase for the Indian forestry debate, the "politics of partnership." For collaboration, even where it does exist, takes place on terms set down by the state, through the officials of the Forest Department. We need to move on to a more inclusively democratic structure, where the state listens to and learns from the community, and where the community itself recognizes and deals fairly with the inequities within its own ranks.

The evidence suggests that contemporary advocates of decentralized forestry have had far greater success than their precursors. One reason for this is the altered political context: Brandis, Pant, and company worked under a colonial, authoritarian regime; the partisans of Chipko and similar movements in a democratic system.

The revival of forest protest in the 1970s also coincided with the international environmental debate, which foregrounded the use and abuse of forests worldwide. The work of Indian scholars had, meanwhile, demonstrated with authority that the century-old history of state forestry in India must be reckoned a failure, in both an ecological and social sense. Finally, the problems with government-directed development programmes in much of the Third World had led to an increasing interest in nongovernmental forms of management and control.

These calls for forest reform from the outside were complemented by pressures from change from within. Starting with West Bengal, the governors themselves, namely the forest officials in charge of their vast landed estate, realized that old methods of control and exclusion were merely fuelling social conflict. An overworked and underfunded bureaucracy then started, slowly, to involve communities in forest working. What started as a strategic imperative became, at least for some forest officers, a sincere change of heart.

Once the critics from without were being echoed by the dissidents from within, the process of reform accelerated. This is indeed the signal lesson of Indian forest history-that meaningful policy change comes about only when the sustained pressure by social movements and their intellectual sympathizers resonates with the feelings of powerful officials within the state bureaucracy.

One or the other, by itself, will not do. When Brandis was active, he was handicapped both by his lone dissident voice within the Forest Department, and by the fact that there had not yet emerged an effective critique from outside. When Elwin, Mira Behn, and others propagated the feelings and aspirations of the peasants and tribals they worked with, the forest bureaucracy was, collectively and to the last man, deaf to their arguments and entreaties.

Forest policy remained unbending and unchanging, with the exception only of the Kumaun hills. There, as we have seen, the popular protests and outside critics were partially successful not because of a honest rethinking by the state but by its concern that this sensitive border region must not be tempted into outright rebellion. Elsewhere, where this political imperative did not come into play, the colonial regime refused to heed the widespread criticisms of its system of forest exploitation.

In more recent times, however, the radical critics have been aided by the autocritique of influential sections of the forest establishment. This confluence of external pressure and internal rethinking explains why, and how, the contemporary proponents of community forestry have, unlike their predecessors, been able to see their ideas and polemic become translated into official policy and (though less assuredly) into official practice. Nonetheless, there are indeed striking parallels between the ideas underlying the application of joint forest management today and the ideas of the early, prescient, and brave but for the most part unheard critics of state forestry discussed in this essay. With respect to the role of forest dependent communities, for example, there is a shared faith in indigenous knowledge, in the management capacity and robustness of local institutions, and above all, a sharp focus on local access to the usufruct of the forests.

Again, with respect to the role of the state, there is a common recognition of the essentially advisory role of the forest department, of its need to collaborate with rather than strictly regulate customary use, and of the justice of sharing revenues from forest working with the villagers.

Then, and now, critics have called strongly for an attitudinal change among state officials, a retraining and retooling in keeping with the democratic spirit of the age. Finally, both past and present proponents of decentralization seem to converge in their larger vision for forest policy in India, a vision which in my understanding consists of three central principles: (1) that benefit sharing (between state and community) and local control are to be the key incentives to ensure sustainable management and minimize conflict; (2) that community-controlled forests would work as a complement to a network of more strictly protected areas, further from habitations, that continue under more direct state control; (3) and finally, that the restricting of state control to these latter areas is vital on grounds of equity (*i.e.*, the respect for local rights and demands), efficiency (*i.e.*, as the most feasible course, with the state not biting off more than it can chew), and stability (*i.e.*, as the most likely way to lessen conflict).

There is little question that the ongoing attempts at reversing or mitigating state monopoly over forest ownership and management do constitute a significant departure from past trends. In a deeper sense, however, contemporary attempts at fostering participatory systems of forest management hark back to a much older tradition. In the late twentieth century, as in the late nineteenth century, there has arisen a movement for the democratization of forest management, for a system founded not on mutual antagonism but on genuine partnership between state and citizen. The first inspector general's vision for Indian forestry was abruptly cast aside in the 1860s and 1870s, but it may yet come to prevail.

3

Forest Estate and Conservation

NATIVE FOREST ESTATE

It is stating the obvious to note that the native forest estate is increasingly the centre of considerable public controversy. The debate over forests is a microcosm of the broader debate surrounding environmental issues because it continually confronts society with complex choices. Each choice offers a variety of paths to the future, and in turn, each path provides net 'value' to society. Net 'Value' is comprised of a mix of quantities and types of 'values' which individually and collectively meet the needs of society. Differing 'paths to the future' confront society with the need to estimate the net value likely to be derived from each route. In economic analysis, and (it is assumed) society at large, the path which offers the most value to present and future citizens is the preferred choice.

OPTIMISING THE BENEFITS TO THE COMMUNITY

Many complex social choices are decided in the realm of politics with some choices arbitrated on a more technical, arms length basis by designated officials. Choices in environmental policy historically combine technical, 'objective', elements and qualitative, 'subjective', elements. As a result, the choices are subject to a range of pressures that may not beset other types of decisions. In this context, it follows that estimating the value, or total 'economic welfare', of different paths within environmental policy may be more difficult than the simpler financial trade offs common to economic decision making. 'Optimising the benefits to the community' has been identified as a key objective in forest policy. In order to achieve this goal, it is necessary to identify the contribution that different policy options can make to the welfare of the community. In other words, we must determine which path to the future offers the greatest 'net value', and which combination of values acts to create that path. This discussion examines the relationship between financial values derived from timber harvesting, the financial values derived from alternative uses of the forest (eg. water, tourism etc), the financial values derived from plantation timber and the non financial values of the forest (eg. environmental amenity) through the reviewing of a series of reports on native forests.

THE CONTRIBUTION OF ECONOMIC THEORY

The Ecologically Sustainable Development Working Group on Forest Use specified three main requirements for sustainable forest use:

- maintaining the ecological processes with in forests (the formation of soil, energy flows, and the carbon, nutrient and water cycles);
- maintaining the biological diversity of forests;
- and optimising the benefits to the community from all uses of forests within ecological constraints. The National Forest Policy Statement adopts these principles as the basis for ecologically sustainable development.

Economic theory sets a standard by which the outcomes are measured, and informs us as to what are 'good' and 'bad' ways of making choices. Economic theory also provides an important underlying philosophy which specifies that changes should be to the net benefit of society at large.

The Role of Economic Theory

Economic theory is often seen as a method, or science, of analysing impacts on the financial economy. When properly and rigorously deployed, however, it is elevated to the 'art ' of choice. Thorough economic analysis concerns itself with how to make choices that provide the greatest net benefit to the community, where 'benefit' implies consideration of all social values. It is for this reason that economic theory, incorporating such concepts as externalities and public goods, has come to play a significant role in decision making processes. The following sections briefly outline the relationship between the concepts of externalities, economic value and economic decision making.

Externalities and the Welfare of Society

Economics is centred on the allocation of scarce resources to competing ends, in order to achieve the most value for human society. In this instance, economic analysis seeks to identify the flows of value that can be derived from 'uses' and 'non uses' of the forest. This process of analysis seeks the combination of uses and non uses that most meets the value preferences of society, both now and in the future. The process is intended to be value neutral in the sense that it is based on the given or expected values of society-whatever those values may be. Traditionally, economic theory uses the 'market' as its benchmark of efficient resource allocation; that is, an allocation of resources that creates the greatest benefit for society. This 'pure' model, however, admits that its results hold true only under very tight and specific conditions. If any of these conditions are breached, then there can be no presumption that the free market outcome is the best one. Externalities arise when certain conditions of a free market do not hold. They arise when the free market fails to include, or take account of goods and services provided incidentally by the completion of another transaction. An externality can be compared to an auction where the main transaction is completed between two parties, but where additional goods and services, not provided for in the original transaction, are subsequently delivered.

These can be delivered either to the main parties or to third parties who were not involved in the original deal, and who, indeed, may even be unaware of the original transaction. For example, if a tree is cut down in the forest for timber, the loss of that tree may adversely affect the view of a home owner, who subsequently experiences a loss in his/her property value.

Externalities and public goods are both expressions of market failure. 'Public goods' refers to a value from whose benefit others cannot be excluded. Thus, deriving pleasure from knowing of the existence of a forest is an externality, because others cannot be excluded from deriving the same pleasure. At the same time, the phenomenon is also a public good because it is non rival in consumption; that is, many can consume the item without reducing the consumption of others. In order to avoid confusion, all such values are called 'non financial values' since they cannot, by definition, be traded in a market place, and hence be expressed in terms of financial value. Where environmental issues are concerned, the market cannot be relied upon to produce an optimal allocation of resources-ie. the greatest overall benefit to the community-because there is no venue, no 'global auction room', in which the community (in particular future generations) can express its preferences over the items it values.

The 'items of value' in this discussion are the 'values' or 'benefits' that flow from the 'use' of 'objects'. Normally, economic analysis talks of objects such as goods and services when the real interest is in the values that flow from these objects. It makes little sense to talk, for instance, of a 'forest' per se. Are we referring to the wood alone, the biological species that reside there, the pleasure that bushwalkers derive, or its function in absorbing greenhouse gases? The term clearly has no operational meaning unless we refer explicitly to its various use values and benefits. In order to avoid any ambiguity, the remainder of this chapter will generally use the term 'values' rather than 'goods' or 'products' (as 'values' is a more inclusive term).

An understanding of externalities is central to quality economic decision making, particularly in the context of the environment-simply because in the real world, there are many items of value (fresh air, knowing that an old growth forest exists, etc) that cannot be bought and sold in the market. The exclusion of these items from the 'economic equation' means that the overall equation is miscalculated. Miscalculations of the total economic equation generally lead to a misallocation of economic resources, and to an overall loss to society. The practical significance of identifying externalities and public commodities relates to their impact on markets. It is impossible for a market alone to systematically produce the socially optimal outcome when externalities and public commodities are present, as they are in the case of the natural environment 1.

PRICING THEORY AND THE FINANCIAL VALUES OF THE FOREST

In economic theory, the price of an item can represent one of three circumstances for the owner.

- The price could reflect the value of the item to the owner.
- The price could value the item at a level greater than the owner's subjective valuation. In this case the owner is likely to sell the item.
- Alternatively, the price of the item may be less than the value to the owner in which case the owner will refuse to sell.

The price of wood must also reflect the different type of values generated by the wood and the portion of forest which is altered by its extraction.

Hence, if the 'public citizen' 2 offers a logging coupe for sale, by definition the value anticipated from the sale must at least exceed the value expected from any alternative use. And additionally, the price needs to equal:

- the costs of the production of wood;
- plus a return on the capital invested, with sufficient reward to cover the risk; and
- total compensation for all non financial values (NFV's) lost in the process of an extraction of the timber.

This pricing method ensures that timber with values exceeding the market price will not be sold. However, the significance of NFV's is that they make up a considerable proportion of the value of timber. When NFV's are destroyed, the loss of value is not felt by one person alone, but is likely to be felt by every Australian citizen, every international citizen and everyone yet to be born. When logging occurs, each of these people experiences a change in the value they derive from the forest, and hence each should be compensated. The price of timber needs to be able to compensate all losers if policy is to be socially optimal. The pricing of timber is crucial to determining the pattern of logging in the forest. In order to implement a solution, the present approach of forestry commissions uses regulation to separate logging areas from non logging areas. This approach could be assuming one of two things. Firstly, it could be assuming that royalties are sufficient compensation for the loss of NFV's 3. In this case, it remains to be demonstrated by those who set royalty levels that they have evaluated the worth of the lost NFV's, and thus set royalty levels accordingly. In the absence of information about how NFV's are determined, it could be assumed that non financial values in the logged areas are considered insignificant to the extent that the Australian community will not require compensation for the losses in NFV's which result. In theory, logging will then occur, in the areas set aside, if the cost of production and capital makes it a financially rational exercise.

Pricing policy is, thus, conducted on a 'compartmentalised basis'. In other words the forest is surveyed for its values, and those areas where NFV's are deemed 'small enough' to avoid the need for compensation, are made available for logging. The remainder is put aside for conservation. Thus, if we use a 'compartmentalised' approach, timber is valued in a two stage process. The forest is allocated to different uses depending on the scale of NFV's in different areas. Forest that is to be subject to logging is then priced on a commercial basis. Pricing takes place on the assumption that the coupes have already been surveyed, and the NFV's have been found to be minor; ie very close to zero.

The worth of the royalties, of course, must also cover the costs of wood production, the return on capital required, and the financial opportunity cost of foregone uses. The Victorian Auditor General's report (1993) indicated the likelihood that the forestry operations in Victoria were making low profits (if any at all). Given that royalties must cover operating, capital and opportunity costs, this leaves very little room to compensate for lost NFV's. In such a circumstance, it is feasible that NFV's are being valued at a worth very close to zero. No evidence was identified to justify such a valuation.

THE THEORY OF TOTAL ECONOMIC VALUE

In any public decision over the net benefits of forests, externalities and public commodities, along with project and alternative financial values, need to be taken into account, as part of the value provided by a 'resource'. In analysing different resource options, a total listing and weighting of values, both private and 'externalised', must be provided for each option. When an economist is comparing alternative social arrangements, the proper procedure is to compare the total social product yielded by these different arrangements... In devising and choosing between social arrangements we should have regard to the total effect. This, above all, is the change in approach which I am advocating.

Coase refers to 'total social product' which we have renamed 'total economic value' in the tradition established by Pearce (1989). In order to compare the total economic value of differing policy options, it is necessary to aggregate for each option all the values (financial, aesthetic, spiritual, etc.) that would be produced for the community if a particular policy option were to be undertaken.

TRADITIONAL PROJECT EVALUATION

Traditional project evaluation normally compares one option against another option, and chooses that which has the most value. In the presence of market imperfections such as externalities, the total economic value (the project's total costs and benefits to society) can be severely miscalculated in at least two ways. First, proper economic evaluation needs to consider the financial impact that one project may have on other parties, not simply whether it is profitable when considered in isolation. Suppose, for example, that the government is requested to give its approval for the establishment of a factory on a river bank. Viewed in isolation, the proposal is attractive: it is expected to provide jobs for 100 local unemployed young people and will contribute an extra $5 million to tax revenues. Viewed from a wider perspective, however, the project's net financial value would be seen to be negative if the poisonous effluent it emits cause the bankruptcy of downstream fish farms and tourist resorts which together generate 200 jobs and $10 million in tax revenues. These 'down stream' financial values are often overlooked by traditional project evaluation methodologies.

The second weakness of the traditional evaluation approach is that it gives little consideration to non financial values. To continue with the illustrative example above, the factory's pollution may poison wildlife and household pets, ruin a picnic spot favoured by thousands of local residents, etc.

These types of costs are real and relevant, notwithstanding the fact that they are of a 'non financial' nature. Resolving the first problem is easier, since it may be possible to obtain the relevant financial data if appropriate economic modelling is conducted. The second problem is more difficult because non financial values are not traded in markets. In other words, there is no 'objective dollar price' which can provide a yardstick with which to compare non financial and financial values. Environmental economists have been working on providing techniques to evaluate non financial values.

Economic Decision Making

As a society, and as individuals, we are constantly evaluating and rejecting options. A social cost benefit analysis is, in effect, the act of making choices. Social cost benefit analysis becomes a conceptual framework that expresses the less formal processes we use to make choices. While the National Forest Policy Statement leaves open the practical question of assessment, it identifies the need for evaluation of policy options. In Diagram 1 we have listed all the values, in the tradition of Pearce (1989) and Young (1992) that appear to be provided by the native forest. Type of value, and where relevant, their relationship to each other. Total economic value (TEV) analysis would require that each policy option would receive a 'score' in each cell of an evaluation matrix. Scores are aggregated by a means yet to be determined which then enables decisions over differing options to be made. Economic decision making seeks to choose between the differing options on the basis of determining that which provides the greatest net social benefit. The means of deciding which option has the greatest net social benefit where 'scores' cannot be aggregated into a single numeraire is, of course, highly contentious.

There are three channels through which resources can be allocated. These are the markets, institutional processes and executive decision making. Markets, however, as decision making institutions, are largely unavailable in the circumstances covered in this report because we are dealing specifically with instances of market failure.

However, modified markets (ie. markets subject to government intervention) are included as a means of policy implementation. The choice between government modified markets and 'pure' market measures should be justified on the grounds of net social benefits.

The irony is that if market measures are to be chosen as the decision making process, then it is a decision that will have to be made by government anyway, and hence does not relieve the public decision making process of the need to evaluate the optimality of value flows from the forests under differing policy options.

In other words, choosing to 'do nothing' is as much a decision as choosing to 'do something'. The lesson is that if optimal social benefits are sought, then market failure automatically triggers government decision making processes-even if that results in a decision to 'do nothing'.

The creation of a formalised process for considering environmental values and other externalities requires a clear understanding of the integrated structure of externalities, economic value and economic decision making. All values-financial, ethical, ecological, etc-must be accounted for, if different policy options are to be compared on a equal basis. The different values that comprise each policy option, need to be integrated in some manner which enables socially optimal choices to be made.

APPRAISING THE VALUES OF THE FOREST

On forests for their views on the appraisal of forest values. Each report was reviewed in order to gain an understanding of the value assessment problems in the forest sector. The reports were examined on the basis of four criteria:

- Identification of externalities.
- Identification of valuation methodologies.
- Identification of pricing/financial valuation practices.
- Identification of likely assessment frame works.

Problems in the Identification of Externalities

A representative example, of the values identified, is listed below. These values were derived from the Resource Assessment Commission (RAC) study. This study, amongst those examined, was the most comprehensive and thorough in its treatment of the issues. The weaknesses in its study were repeated, and magnified, in other studies.

Assessment Commission Inquiry into Forest and Timber

Conservation; wilderness; recreation; Financial values such as wood, tourism and water supply, oils, seeds, tannins, medicinal plants, sandalwood, bark products, fishing, hunting, wild foods, infrastructure corridors; Information via education, research; intrinsic values like aesthetic (different from visual), heritage, spiritual, biocentric, wilderness, lifestyle; Utilitarian values; 'pristinity'; landscape; scenic quality, public sensitivity, industry value by implication; existence values, option values, vicarious use values, bequest values, ecological sustainability, ethical values.

The main weaknesses in assessing the manner in which values had been surveyed were that:

- Values were not identified or listed in a systematic or comprehensive manner. The above extract from the RAC report was amongst the most comprehensive, but even so, lacked a systematic listing of values.
- Different value categories often overlapped. For example, ecological sustainability was repeatedly confused with 'environmental amenity'.
- There was also repeated confusion over the values we impute to an object, and the values we derive from that object as a result of our preferences. For instance, a forest site may have great scientific value, but not be of great aesthetic value. Conversely, a forest site may have limited scientific value, but great aesthetic value. In the latter case, a purely scientific view point would not reveal the aesthetic importance of the site. Similarly, there are other values-for example existence value-held subjectively by the public with respect to native forests. A study which evaluated objective characteristics only would not pick up the subjectively felt values. We have identified this as the 'object subject' tension.

Confusion over the type and source of 'values' will undermine the efforts to assess all values. It is therefore necessary to have an agreed list of all values, comprehensively and systematically ordered, in order to achieve the goal of optimising benefits to the community. The twentynine uses identified each provide seven different types of value which give their total worth. For example, use D 11 (craft industries) provides D 11.1 (direct use value), D 11.2 (option value), D 11.3 (quasi-option value), D 11.4 (existence value-sometimes known as industry value), D 11.5 (intrinsic value), D 11.6 (combination value) and D 11.7 (bequest value). Each of these seven components combines to give the total value provided by craft industries. In turn, the 'use' known as 'craft industries' can be combined with a package of other uses that will generate the total economic value of the forest.

While within a particular use, different components can be combined to provide a total value for that use-this does not imply that the relationships between each component are necessarily positive. They may in fact be inverse for a range of values. Similarly, the relationships between different uses can be positive, negative or zero-even though total economic value is additive of the components.

Identifying Suggested Valuation Techniques

Each report suggested, implicitly or explicitly, a range of valuation techniques for providing an 'objective' assessment of the values identified. Overall, the lack of comprehensive identification of values, common to all thirteen reports, subsequently inhibited the matching of valuation techniques to appropriate categories of value.

This complicated the process of value assessment since it provided limited guidance about how various values should be measured.

On the positive side however, the Resource Assessment Commission report did provide an assessment of each evaluation technique, which marks a very useful first step towards providing a listing of techniques relevant to the assessment of each value.

The Organisation for Economic Co-operation and Development references (OECD, 1986; 1992) also made a worthwhile contribution towards evaluating techniques.

In general, the reports provided only limited advice on the means by which different types of values should be evaluated. Moreover, with the exception of the RAC and the OECD reports, the techniques put forward were generally insufficient to measure the full range of values derived from the forest.

For example, the reports would discuss techniques for measuring the scientific values of the forest, with out discussing other measures that evaluate the 'worth' of the non-scientific values.

Whilst some discussion of uncertainty was entered into by four reports, the issue was not raised sufficiently to provide guidance on how valuation techniques were to account for uncertainty. What is required is a more explicit discussion on the means by which 'uncertainty' should be handled in valuation processes.

The valuation techniques that were identified. While there will be some debate about whether these are all 'techniques' or 'management strategies', it is clear that the implementation of these processes leads to the de facto valuation of forests.

For the purposes of this report, we have described them all as valuation techniques. Some of the techniques identified below were implicit in the various reports rather than explicitly described.

THE ASSESSMENT FRAMEWORK AND FOREST POLICY

Reviewing the various reports on the issue of forest policy has provided an insight into the process by which decisions are made. To summarise the inadequacies that have been identified in the respective assessment frameworks and processes.

Assessment Frameworks and Processes

There are several assessment frameworks used in Australia for forest decision-making. Most of the reports have only considered assessment frameworks in general terms, without specific reference to real examples. In order to evaluate these assessment frameworks, the report provides a 'checklist'. The 'checklist' provides a comprehensive series of questions which any worthwhile assessment framework should be able to answer. The 'checklist' approach facilitates identification of the broad strengths and weaknesses of existing or proposed assessment frameworks

A common theme in all reports (both Australian and international) was an underlying philosophical assumption that a 'technical', non-political choice process was possible, and to some extent desirable. This process was not challenged explicitly in any report, despite the obvious claim that the political arena is the desired place for settling value disputes (indeed it could be argued that this is the raison d'etre of politics). For the moment, however, this discussion accepts the premise that a 'technical' or 'black box' approach is feasible for value choices. So as to gain an understanding of international experience, the report has sought to review processes suggested by the OECD and the World Bank. These internationally-recommended approaches provide a benchmark for achieving 'world best practice' in this field. The report has also examined the type of system deployed by New Zealand under the auspices of the Resource Management Act.

The international reports reflect a worldwide concern over the assessment of environmental values. While it is now recognised that environmental values must be incorporated, it is not known how this should be achieved. The World Bank view is not unusual:

Externalities are thus clearly troublesome, and there is no altogether satisfactory way to deal with them. This is no reason simply to ignore them, however; an attempt should always be made to identify them and, if they appear significant, to measure them. This accurate, but unhelpful commentary is largely due to a neglect of environmental factors within economic analysis. This was noted by two internationally renowned economists in 1988, with respect to the field of development economics:

Environmental resources appear in this literature about as frequently as rain falls on the Sahara. They also noted, as recently as 1988, that one of the world's best known texts on development economics... simply has no discussion of environmental resources and their possible bearing on the development processes.'

This is symptomatic of neglect within the broad field of economic analysis. The OECD (1992) remarked that conventional cost benefit analysis would not include a number of environmental effects ...because of the lack of market price data or easily accessible direct proxy price data.'

Given this neglect the OECD report (1986) on the public management of forests, must be regarded as a landmark publication amongst the reports reviewed in this study. Overall, given that all economic activity centres on the idea of 'development', the general lack of incorporation of environmental issues is a serious omission. As a result the weaknesses identified in Australian assessment processes are not unusual when compared with 'world best practice'-though this does not make such deficiencies acceptable according to economic theory. Neo-classical economic theory would state quite clearly that on a worldwide basis, as well as locally, such an omission must be regarded as a threat to the economic well-being of society.

Existing Assessment Frameworks

Australia, with its federal political structure, has many different means by which choices over values are made. Rather than describe this diverse array of assessment frameworks, this book will provide a thumbnail sketch of the key features of assessment frameworks and processes. Along with the international examples cited, seem to point to three key characteristics of concern in any assessment: 'Who makes the decisions' is significant because it determines whose values are brought to the process. The economics profession has long argued that institutions tend to be captured by the values of specific groups, and that to expect any public institution to be value-free and working purely in the public interest is naive. From the perspective of assessment processes, it is important that advocates of differing values are given an equal voice in the process of analysis and decision-making.

The RAC Forest and Timber Inquiry report took some interest in the Victorian model which involved the Land Conservation Council in decisions about selection of forest logging coupes. The AHC-CALM model of joint survey work also received some favourable attention, though its joint deliberations were designed to identify values rather than make decisions. That no other systems were singled out for favourable comment, in the various reports reviewed in this study, seems to indicate that other processes may be lacking an adequate contribution from representatives of other values.

It is not unfeasible that an appropriately structured institutional process could make optimising decisions-given that the right information on values was available. The information provided, as outlined by the thirteen reports reviewed in this process, is inadequate for the task of making choices over values. The AHC-CALM model received some favourable comments for its process from the ESD and RAC reports Certainly, its assessment method is systematic and thorough. Its prime weakness is that it does not consider, in its own words, 'social values'. 'Social values' as defined by the AHC, appear to cover that group of values described by economists as existence value. As a result, it only acknowledges some of the values considered important in a total economic assessment. That it was commended as a model for others suggests that even analysis of 'non-social' values is deficient in existing assessment frameworks and processes.

Other reports such as the Victorian Auditor-General's report and the NSW Public Accounts Committee confirm the impression that only poor data is available on values.

Financial values are also a significant part of the analysis. A significant range of reports located deficiencies or made recommendations that pointed to opportunities to improve the gathering of financial data. The deficiencies in this regard extended to the data on both financial values generated by forestry and financial values generated by other activities. The existing assessment framework and processes were not being supplied with the relevant data because the necessary mechanisms such as appropriate accounting systems are not adequate.

Compare Different Values

Assessment frameworks must make choices between values. Few technical models were offered on how such choices should be made,outside the political process. The Victorian Auditor-General's report noted that the previous system in that State had used a committee of relevant officials, backed up by a requirement to find options, with a final 'safety valve' being provided by there being recourse to the Minister. No doubt similar methods are used elsewhere, though, what is required is a method that considers all values at the point of decision with appropriate systemic checks and balances.

The AHC-CALM model and the process offered by the RAC Forest and Timber inquiry were the closest Australian examples of technical approaches which are beginning to seek full consideration of all values. The OECD (1986) report is a benchmark in this field. Only the process implied in the RAC report comes close to this analysis. It is assumed that if there were another system, the RAC inquiry would have located it. The existing assessment frameworks, obviously, can be said to rely on the judgements of those who take the decisions and the information that they have been supplied with. There must be serious doubts that these existing frameworks have been properly structured. A assessment framework and process needs to specify who decides, with what value information and by what intervalue comparative techniques.

Inadequacies in Pricing/financial Valuation

The significance of non-financial values to the calculation of financial values is considerable. The forest agencies and all the reports reviewed in this study have adopted a compartmentalised approach to pricing. The compartmentalised approach assumes that assessment of choices between values occurred prior to the pricing of timber. If non-financial values are improperly assessed, then the resulting effect is that financial values will also be inappropriately founded. In this sense a compartmentalised approach to pricing is a practical manner of settling commercial pricing issues. In simple terms, if non-financial values are being compromised it is due to the lack of adequate assessment prior to the commercial pricing process.

PRICES OF FOREST TIMBER

Pricing of logs should, however, be offered on a normal commercial basis. It is fairly clear from the reports that have been reviewed that operations are not presently proceeding on a commercial basis.

Prices of forest timber are being distorted by a range of factors that are, in general terms providing a subsidy to the logging of native timber, whilst other factors are raising the costs of private plantations. Other factors such as the lack of rational markets for the pricing of water and other services, partly or wholly derived from forests, are raising the relative cost of alternative financial values.

The following distortions in commercial pricing of timber have been identified. The list has been 'cobbled' together from all the reports reviewed in this survey.

1. The lack of marginal cost pricing.
2. The non-existence of effective commercial accounting systems.
3. Long-distance transport subsidies.
4. Non-payment of rental for use of public land.
5. Low royalties; ie. low log prices for timber
6. Possible cross-subsidisation from community service obligations.
7. No liability for payment of a resource rental tax.
8. No liability for payment of a notional income tax (presumably meaning company tax).
9. No liability for local government rates.
10. No liability for sales tax on vehicles, plant and equipment.
11. Non, or low, payments of interest on borrowed capital.
12. Non-payment of dividends.
13. Low user charges for cost recovery from long-distance transport
14. The lack of commercial discipline due to lack of a corporatised commercial structure;
15. Serious deficiencies in the tax system for plantations;
16. Problems with planning, rating and exporting from plantations;
17. Problems with the small numbers of buyers and sellers of timber;
18. Problems with the purchaser behaviour of large buyers of timber;
19. The dominant market role of state forestry agencies;
20. Failure to properly account for non-financial values in pre-logging operations either by exclusion of areas or by forestry codes of practice.

21. Failure to adequately assess the potential financial contribution from sales of goods and services such as water from the native forest.
22. Failure to separate commercial operations from regulatory functions.

They also felt that the market price was too low and that international prices were not as significant in pricing native forest timber as was considered by the Commission. They felt that the native forest timber price should be raised to 'the point of indifference.' (?) and that the case for doing so was 'overwhelming'. This is even more significant when it is considered that 'the Forestry Commission sets these prices...the Commission is clearly a price leader.'

Inadequacies in Assessment Frameworks

The existing assessment framework is in a period of transition. The post-1945 charter of the forest agencies to push for economic development has changed. The capital resources of government and the Australian economy are scarce and need to be deployed for maximum financial benefit or for some clearly identified social 'good'. Changes in social values are also placing pressure on the old system. These two forces, in combination, are leading to greater scrutiny of forest activities, and demanding greater levels of accountability. It is clear that assessment processes are lagging in their response to these issues.

Entrepreneurial Capture

The existing assessment process is vulnerable to 'capture' by entrepreneurs proposing one-o ff developments. This limits government ability to consider a range of project options as proposed by the OECD (1986) and the World Bank (1992) in particular. The existing Australian approach of having a single project option placed in front of decision-makers must be regarded as inferior, if not a structural distortion of the decision-making process.

Lack of Commercially based Accounting of Projects

The lack of commercially-based accounting data for project analysis means that it is not possible to accurately evaluate true project financial worth. This means that many unprofitable activities are slipping through the policy net adding to the economic burden already being borne by Australians. The effect of such methods is to inflate the relative worth of financial values vis a vis non-financial values. It also inflates the relative values of forestry operations vis a vis the alternative activities such as plantations and water production.

NFV's are not being Assessed

The most serious flaw is that it appears that assessment of non-financial values is simply not occurring. The concentration of reports on discussing aspects of how such analysis could be done, rather than a focus on how it is being done leads to the conclusion that serious analysis of the relative worth of non-financial values is yet to begin. It would seem fair comment that if relatively simple choices between wood and water production are not being evaluated, then other more complex interactions are also being neglected.

Lack of a Comprehensive set of Values Descriptions

The next major flaw in assessment processes is the lack of a complete set of values relating to forestry. Each report examined has mentioned a range of words that seek to describe, in particular, the non-financial values, but there has not been a report which systematically listed what society derives from the forest. It is little wonder then that assessment frameworks barely exist or function if the values are not even identified.

Lack of Focus on the Full Suite of Values

Where non-financial values are considered, they tend to relate only to the assessment of physical characteristics of the forests. All reports acknowledge the need to protect these physical characteristics, for example 'sustainable yield', yet there is evidence that they are not being protected, even at this rudimentary level (. If physical characteristics are only barely being protected, then it is reasonable to assume that other more abstract and intangible values of the forest are not being taken into account. The Dorrigo Management Area Environmental Impact Statement does not mention any non-financial values other than soil, birds, animals and recreation. Ecological sustainability and existence values are barely considered at all.

Public Involvement: Preached by Some but not Practised

The practice of assessment seems to place too little stress on the need to obtain public review and input into forestry decisions. In an area where clashes over values are common, the need for representation by different values should be considered. As a result the political process is starved of information about social preferences. The OECD (1986) reveals a considerable commitment to public involvement, in particular their example of the 'Inner Valley Road Project'.

Future Generations Considered by Implication Only

The OECD (1986) report was not without its flaws. Its assessment framework gave little attention to considering the impact of future generations from any valuation assessment process. In economic theory it is impossible to internalise the interests of future generations simply by assessing the preferences of the present generation. Neither market processes or valuation techniques, in the face of irreversibility, are capable of 'looking after' the interest of future generations in any manner that can be relied upon. Fortunately the concept of ecological sustainability goes some way to achieving this goal. The internalisation of the interests of future generations requires explicit discussion.

Most reports steered clear of identifying procedures that would resolve the choices thrown up in the forestry debate. The RAC report (1992) went furthest in this regard but tended to educate by demonstration rather than explicit prescription. The ESD report (1991) had some brief comments, but if the assessment of these different values is to proceed satisfactorily, it is clear that the hard issue of 'assessment frameworks' must be dealt with honestly.

DIRECT EVALUATIONS OF NON-FINANCIAL VALUES

The OECD indicated that other nations are moving towards direct evaluations of non-financial values. Australia's approach would appear to be somewhat slower than average, though of course there are some institutions such as the World Bank that have acknowledged the need for change but are yet to make the practical changes that are required.

It was significant that the United States was acknowledged as having greater experience in the area of non-monetary evaluation. The experience was deemed to be due to the 1969 Environmental Policy Act which required quantification in non-monetary terms of all environmental impacts. The OECD concluded that it would seem that education and more valuation projects are a prerequisite for more effective utilisation of the techniques. A learning process is needed..

Valuation techniques, whether monetary or non-monetary, form the backbone of an assessment framework. The same learning process is required for assessment processes as well. Many OECD countries are learning but Australia does not appear to have started.

Assessing the Assessment Frameworks

A major difficulty with evaluating an assessment framework is the sheer scale of the task itself, as anyone who has read an Environmental Impact Statement will testify. The result is a tendency to focus on one or two key angles to the exclusion of other angles. To provide a systematic 'framework' for assessing assessment frameworks. A series of questions is provided, that an assessment framework, if it is to be successful, should be able to address. The existing Australian and international assessment practices are evaluated has been tested on actual reports, in this case the Victorian Auditor General's report as well as an analysis of the Dorrigo Management Area Environmental Impact Statement.

The native forest is capable of producing three major categories of values. These are:

- financial values from timber production;
- alternative financial values from other forms of forest use; and
- non-financial values.

This report has sought to examine the issue of giving each type of value 'full and due consideration' in policy processes. Three themes have recurred throughout this study.

- There is a theme focusing on the analysis of technical comparisons between different values. That is the use and abuse of various techniques that quantify values.

The native forest is capable of producing three major categories of values. These are:

- financial values from timber production;
- alternative financial values from other forms of forest use; and
- non-financial values.

This report has sought to examine the issue of giving each type of value 'full and due consideration' in policy processes. Three themes have recurred throughout this study.

- There is a theme focusing on the analysis of technical comparisons between different values. That is the use and abuse of various techniques that quantify values.
- Asecond theme revolves around the nature of the institutional arrangements that ensure value selection which is socially optimal.
- The major theme, however, is achieving a level playing field between the major types of values derived from the forest.

The emphasis on achieving a level playing field is in the spirit of the microeconomic reform process that has taken place over recent years, and continues with recent reports such as the Hilmer Inquiry. Sims described microeconomic reform as occurring when changes are made to achieve more output from a given level of inputs.. The output from the forests is intended to maximise the social welfare of the community. For this reason, microeconomic reform in the forest sector must seek to arrange the three sets of values above so that welfare is maximised.

NEED FOR MICRO-ECONOMIC REFORM

The implication of the need for micro-economic reform is that the sector is not making an optimal contribution to the quality of Australian life. This review of reports on the forest sector suggest that there is considerable potential for microeconomic reform in two dimensions.

- In broad social terms it would appear that the community is not maximising the social benefit from the forest resource.
- Secondly, within the social benefit the community receives, the financial value to be had from the forest resource is not being maximised, assuming that all other types of values remain unchanged (ceteris paribus 'All other things being equal.').

If the microeconomic reform process (optimising social benefits) in the forest sector is to succeed, the estimation of the worth of the three categories of value described above must be accurate. It would appear that estimation of these three categories of value is presently inaccurate. This inaccuracy would confound any attempt to derive optimal social benefits.

IMPACT OF MISCALCULATING FINANCIAL VALUES

The impact of miscalculating the financial value of timber production is significant for Australia's macro-economic performance. Firstly, it overestimates the financial worth of timber production vis a vis some other form of commercial activity. The Read Sturgess report (1992) made it clear that it was more financially valuable for the Upper Thomson catchment to supply water than to supply timber, and consequently more profitable for the economy. The overestimation of the worth of forestry would tend to hide this potential economic 'free lunch'. Other competing uses for the forest may be similarly affected. More generally, the over-valuation of forestry financial values disguises what may be net losses in this sector which are actually diminishing the wealth of Australia unlike other commercially viable industries. Failure to accurately value forestry also leaves the commercial operations of the forestry commissions vulnerable to an income shortfall in the event of corporatisation. Even without corporatisation, the poor measurement processes for forestry blind the commissions to other profit centres such as leasing forests to supply water to regional communities with expanding industries.

Over-valuation of forestry through poor input pricing, in addition to affecting competing uses of the forest, also undermines industries that provide substitutes for forest timber ranging from plantations to steel production. Inefficient input pricing in native timber production undermines all competing industries, all other things being equal. This report has focused on the 'collateral damage' caused to the plantation industry. However, what applies to the private plantation industry, also applies, probably on a lesser scale, to other industries that compete with native forest timber. The private plantation industry is one clearly visible victim of an 'unlevel' playing field in the forest sector.

The over-valuation of native forest harvesting undervalues, by definition, the product of plantations. This leads to lower levels of investment, employment and wealth creation than may otherwise have occurred. It prevents the development of value-added timber industries that can further develop on the basis of a plantation resource. This presents the Australian economy with a 'double loss'. T h e macro-economy loses, marginally at least, from harvesting native forests, and then loses again through the disincentive effects on plantations. Other industries may suffer as well. Finally, there is a third loss built into the playing field, which are the microeconomic distortions, such as poor taxation arrangements, identified by NPAC (1991), that inhibit the plantation industry. The microeconomic distortions in the plantation sector alter the financial calculus such that the value of native timber is enhanced relative to plantation timber. In addition these microeconomic distortions cause a further round of macro-economic losses from the plantation industry. The structural distortions within the forestry sector, and within other sectors both competing and substitute, compound the overvaluation of forest timber relative to other values. This causes a loss of income, jobs and a waste of capital as private agents respond to these inappropriate signals.

The Effects of Miscalculating NFV's

The second major impact of overvaluing forestry financial values, is felt on non-financial values. Over-estimating of forestry financial values also underestimates, by definition, the relative worth of non-financial values. Thus the process of evaluating non-financial values is seriously handicapped from the beginning. The significance of this was pointed out by Sims, in another context, when he stated

Clearly our objectives are broader than maximising GDP. For example pollution in our cities is of general concern. It might be that including externalities in prices would decrease environmental damage and decrease growth, but improve overall living standards.

Two points flow from the comments by Sims. Firstly, Sims appears to be giving emphasis to a commonly-held view that environmental improvement can only be achieved at a cost to the economy. This review has indicated, in a very clear manner, that in the forest sector financial gains can be achieved whilst making environmental improvements. The view espoused by Sims presumes that the 'balance' between environmental and financial considerations is 'efficient'; that is, in the context of this report, the evaluation of the financial worth of native forests has been accurately calculated. The evidence indicates that the balance is inefficient, because calculations of financial and alternative financial values, as well as NFV's are inaccurate. Therefore, this report indicates that improvement in environmental conditions, over a very large margin, will improve financial conditions as well.

At some point along the policy continuum, improving environmental conditions will negatively affect financial conditions-but the forest sector, along with many other sectors, is not at this point, and indeed is a long way from it. Microeconomic reform in the forest sector has the potential to bring environmental improvement in its wake by removing financial inefficiencies that have contributed to over-exploitation of the environment. The implication of financial 'inefficiency' in the sector needs to be clearly understood. It implies that society could be financially better off if certain microeconomic reforms were undertaken. The implication of 'economic' inefficiency is that, in the context of this report, the environment and the financial wealth of society, could be better off, leading to an unambiguous improvement in society's level of 'satisfaction'. In this sense, when 'financial inefficiency' is present, microeconomic reform and environmental policy can be complementary, and not competing policies.

Secondly, failure to evaluate non-financial values relative to financial values accurately is critically important to maximising 'overall living standards'. The benefits that are provided to Australians from NFV's can also be described as 'psychic income'. These benefits also have the fortuitous 'collateral' benefit of also attracting tourists, and hence contributing to the expansion of this critical export industry.

Non-financial values provide direct inputs into the daily consumption of Australian citizens and even to foreigners through tourism, wildlife films and existence values. An assessment framework which seeks to optimise benefits to the community, but which does not recognise these non-financial values can generate financial flows that actually undermine community well-being.

This happens because the source of much 'psychic income' (ie. the natural environment) has been lost and replaced by financial benefits that are too small to compensate.

If non-financial values are to be given their full and due consideration, then financial and alternative financial values must be correctly evaluated. This review has identified that there are significant inefficiencies in the estimation of value in the forest sector. The presence of these inefficiencies in the market activities of the forest sector is due to structural distortions.

FINANCIAL AND ENVIRONMENTAL PERFORMANCE

The structural distortions contribute to poor financial and environmental performance. There is a considerable social, environmental and financial dividend to be achieved by microeconomic reform in the forest sector. Microeconomic reform requires that a 'trilogy' of assessments be made: evaluate financial values (FV's), evaluate alternative financial values (AFV's) and evaluate non-financial values (NFV 's). Continuation of present practices such as failure to 'estimate value' accurately will cost Australia in lost jobs, lower incomes and a diminished environmental quality of life.

Here this report distinguishes between financial gains that contribute to the overall health of the financial economy, and economic gains that contribute to the overall state of social well-being. Improvements in financial well-being are commonly referred to as economic gains, for example improved resource allocation, because of the easy but methodologically sloppy habit of referring to the financial activities of the Australian society as the 'economy'. Accurately the economy is the process by which we gain 'welfare', a part of which is the financial gains offered by financial activity.

In another circumstance, when a Treasury officer was asked to described the non-pecuniary benefits of working in Treasury, he called these benefits 'psychic income'.

THE ECONOMIC DIMENSION OF TRANSITION IN FOREST

One way in which an economist familiar with cost-benefit analysis (CBA) may look at environmental impact assessment (EIA) is that EIA is a CBA requiring a large amount of information rigorously collect-ed and analyzed by a group of scientists of various disciplines. One can appreciate that the development of EIA has served to advance the systematic descrip-tion and quantification of environmental effects in a way which can only serve to improve the quality of CBA undertaken.

One can also appreciate that EIA constitutes a process, formalized through environ-mental legislation, regulations and procedures, which in turn reinforces the community participation re-quired for informed decision-makers seeking the common good. While CBA within EIA is a key gen-erator of information to the decision-makers, recog-nition should also be given that the EIA process is both a planning tool and an execution monitoring tool. Thus, the role of CBA is not as an intermediate step, but rather as an integrated parallel process.

Hence, the CBA should follow up, quantify and eval-uate the dynamic iterations created by the exhaustive search of alternatives, by the efficient inclusion of en-vironmental protection and mitigation measures and by the explicit valuation of preference and choice ar-ticulated by community, society and their representa-tives acting in the best interest of present and future generations.

VALUE OF CONSIDERING ENVIRONMENTAL EFFECTS

The main purpose of the economic analysis of a project is to ascertain whether the project can be ex-pected to create more net benefits than any other, mutually exclusive option, including the option of not doing it. Consideration of alternative options, therefore, is a key feature in proper project analysis. Often, important choices about alternative project options are made early on in the project cycle.

These options may differ considerably in their general eco-nomic contribution, and they may also differ greatly concerning their environmental impact. Therefore, including environmental effects in the early econom-ic analyses, however approximately, should improve the quality of future decisionmaking.

The CBA process-defining objectives, searching for alternatives, costing out the resources involved, speci-fying the effects of each option involved and compar-ing all the costs and benefits-normally requires considerable efforts. In the case of Cuba, we are faced with two joint families of economic valuation prob-lems: the "environmental" ones and those problems associated with the incipient transition from a cen-trally planned economy to a truly participative mar-ket based economic system.

In principle, economic analyses are to take into ac-count all costs and benefits of a project. With regard to environmental impacts, however, there have been two basic problems even in developed countries. First, environmental impacts are often difficult to measure in physical terms. Second, even when im-pacts can be measured in physical terms, valuation in monetary terms can be difficult. In spite of such dif-ficulties, a greater effort needs to be made everywhere to 'internalizes' as many environmental costs and benefits as possible by measuring them in money terms and integrating these values in the economic appraisal. The measurement in money terms are made even more difficult in countries undergoing market reforms. The environmental valuation problems can be sum-marized as follows:

VALUE OF ENVIRONMENTAL ASSETS

While man-made and human capital may be valued with relative ease by observing existing market sys-tems, where available; the existence value of clean wa-ter and air, tropical forests, wetlands, coral reefs and other environmental assets and their functions is much more difficult since not even market prices can reflect their full contribution to other economic ac-tivity and to human welfare. In particular, the mar-ket price of water does not reflect the various services it provides nor do market values can accurately re-flect what happens when irreversible loss or damage of natural resources occurs as environmental degrada-tion exceeds a critical threshold level.

COMPLEXITY OF ENVIRONMENTAL VALUATION

Complexity of environmental valuation also arises due to the multi-ple functions of being a source of raw materials and energy, being a sink for assimilating man-made wastes and providing other services such as recre-ational/tourism services, storage of genetic diversity and scientific and educational benefits.

The following classification is useful:

- Direct use values are derived from the economic uses made of the natural system's resources and services. Examples of these are outputs such as timber, game and recreation from forests or fish and scuba tourism from coral reefs.
- Indirect use values are the indirect support and protection provided to economic activities and property by the resource system's natural func-tions or environmental services. Examples of these are watershed protection and soil erosion prevention provided by forests, and beach sand and mooring facilities protection provided by coral reefs.
- Non-use values lie in the special attributes of the natural system as a whole, its cultural and heri-tage uniqueness; it includes both existence and option values. Existence values reflect public goods which can

be enjoyed by more than one consumer without decreasing the amounts en-joyed by others such as clean air, beaches and forests. The existence value is the utility that consumers derive from just knowing the public good exists. A way of measuring that utility would be to measure the willingness to pay or the contingent value assigned if the public good were to disappear. Option values reflect what current generations wish to bequeath for future generations to inherit. They imply both an ethi-cal commitment to sustainabilty for the children of our children and in a shorter time-frame the maintenance of options to solve current prob-lems. Examples of option values have been devel-oped for forests relating to biodiversity and the search for cures of cancer and AIDS-"if forest were to disappear then the options to find such cures will vanish" or "as long as the forest is protected there is the option to find the cure".

Direct Effects Valued on Conventional Markets Some methods are directly based on market prices or productivity. This is possible where a change in envi-ronmental quality affects actual production or pro-ductive capability.

Change-in-Productivity: Development projects can affect production and productivity positively or neg-atively. The incremental output can be valued by us-ing standard economic prices where available. Loss-of-Earnings. Environmental impacts can signifi-cantly affect human health. In theory, the value of health impacts should be determined by the willing-ness to pay of individuals to maintain their health. In practice, one uses earnings lost upon early death, dis-ease or job absence. This approach is used in highway and industrial safety, and in air pollution studies.

The "implicit value of human life" approach is reject-ed by many as dehumanizing since human life can be said to have infinite value. However, society, govern-ment regulations, insurance companies and judicial courts implicitly and explicitly place finite values on human life and health. This is a necessity reflecting limited resources to be allocated for health expendi-tures. The relatively high level of health expenditures in Cuba would indicate a high implicit value of hu-man life and health. However, one can also discern a political motivation and its attendant benefits behind it.

Preventive Expenditures: Individuals and governments invest in prevention measures to avoid or reduce un-wanted environmental effects. Environmental dam-ages, are often difficult to assess, but historical information on preventive measures and their costs may be interpreted as a minimum value for the expected benefits that the preventive measures seek. If it is found, for example, that there is a historical pattern of under investment in Cuba for natural disaster planning and prevention; then, it can be inferred that the benefits expected from such preventive measures have been very low. This conclusion would in turn imply that the loss of human life would be assigned a relatively low value. While this possible conclusion may appear to conflict with the high political priority for health expenditures, it can be observed that the level of damages or loss of lives caused by natural di-sasters cannot be easily attributed to government in-action and thus the political cost may be easily ex-plained away.

Potential Expenditure

Valued on Conventional Markets

Replacement Cost: Simply, the costs that would have to be incurred in order to replace a damaged asset. The estimate is not a measure of benefit of avoiding damage since the damage costs may be higher or low-er than the replacement cost.

Valuation Using Implicit (or Surrogate) Markets

Sometimes one must use market information indi-rectly. Approaches to be considered are the travel cost method, the property value approach, the wage dif-ferential approach, and uses of marketed goods as surrogates for non-marketed goods. Each technique has its particular advantages and disadvantages, as well as requirements for data and resources. One must determine which techniques might be applica-ble to a particular situation.

Travel Cost: This approach measures the travel cost which reflects the willingness of consumers or users can serve to measure the benefits produced by recreation sites (parks, lakes, forests, wilderness). It can also be used to value "travel time" in projects dealing with fuelwood and water collec-tion.

Property Value: This valuation method is based on the general land value approach and can determine the implicit prices of certain land areas. The property value approach can help analyze willingness to pay for properties with different pollution levels and infer the implicit cost of pollution. The method compares prices of houses in affected areas with equal size and similar neighborhood characteristics elsewhere in the same metropolitan area.

Valuation Using Constructed Markets

Contingent Valuation: When society's preferences as revealed in market prices are not available, the con-tingent valuation method tries to obtain information on individual preferences by posing direct questions about willingness to pay. It basically asks people what they are willing to pay for a benefit, and/or what they are willing to accept by way of alternate compensa-tion to tolerate an environmental cost. This process may be achieved through a direct questionnaire/sur-vey. Willingness to pay is difficult to measure and de-pends on the income level of the sample subjects, and involves problems of designing, implementing and interpreting questionnaires. While its applicability may be limited, there is now considerable experience in evaluating the quality of supply of potable water and electricity services.

Artificial Market: Such markets can be constructed for experimental purposes, to determine consumer willingness to pay for a good or service. For example, a home water purification kit might be marketed at various price levels, or access to a game reserve may be offered on the basis of different admission fees, thereby facilitating the estimation of values placed by individuals on water purity or on recreation facilities.

THE DISCOUNT RATE

Discounting is the process by which costs and bene-fits occurring in different time periods may be com-pared. The discount rate to be used has been a gener-al problem in cost-benefit analysis, but it is particularly important with regard to environmental issues, since some of the associated costs and benefits are very long-term or irreversible in nature. In stan-dard analysis, past costs and benefits are treated as "sunk" and are ignored in decisions about the present and future.

Future costs and benefits are discounted to their equivalent present value and then compared. In theory, in a perfect market, the interest rate re-flects both the subjective rate of time preference (of private individuals) and the rate of productivity of capital.

Higher discount rates may discriminate against fu-ture generations. This is because projects with social costs occurring in the long term and net social bene-fits occurring in the near term, will be favored by higher discount rates. It is often argued that discount rates should be low-ered to reflect long-term environmental concerns and issues of intergenerational equity. However, this would have the drawback that not only would eco-logically sound activities pass the cost-benefit test more frequently, but also a larger number of projects would generally pass the test and the resulting in-crease in investment would lead to additional envi-ronmental stress.

Many environmentalists believe that a zero discount rate should be employed to protect future genera-tions. However, employing a zero discount rate is in-equitable, since it would imply a policy of total cur-rent sacrifice, which runs counter to the proposed aim of eliminating discrimination between time periods-especially when the present contains wide-spread poverty.

In the case of projects leading to irreversible damage (*e.g.*, destruction of natural habitats, etc.), the bene-fits of preservation may be incorporated into stan-dard cost-benefit methodology using the Krutilla-Fisher approach. Benefits of preservation will grow over time as the supply of scarce environmental re-sources decreases, demand (fueled by population growth) increases, and possibly, existence value in-creases. The Krutilla-Fisher approach incorporates these increasing benefits of preservation by including preservation benefits foregone within project costs. The benefits are shown to increase through time by the use of a rate of annual growth. While this ap-proach has the same effect on the overall CBA as low-ering discount rates, it avoids the problem of distort-ed resource allocation caused by arbitrarily manipulating discount rates.

CONCLUSIONS

In order to achieve economically sustainable manage-ment of natural resources and environmental protec-tion, one must effectively incorporate environmental concerns into decision making through the EIA pro-cess.

This presentation has reviewed concepts and tech-niques for economic valuation of environmental im-pacts within EIA procedures. The process of internal-izing these environmental externalities can be facilitated by making rough qualitative assessments early on in the project evaluation cycle-the advan-tages of which would include:

- early exclusion of alternatives that are not sound from an environmental point of view;
- more effective in-depth consideration of those al-ternatives that are preferable from the environ-mental viewpoint; and
- better opportunities for redesigning projects and policies to achieve sustainable development goals.

In order to fully reflect society's values and preferenc-es about environmental values, non-market methods of estimation can be used and will enhance commu-nity participation through well designed and admin-istered questionnaires and surveys. Research and training about EIA and embedded economic analysis methods is needed in Cuba. As developing countries learn to successfully apply EIA methodologies the goals of sustainable development will become more attainable.

4

Forest Management

Forest management expenditures, which are made in the expectation of creating benefits at some time in the future, are best viewed as investments. Although some foresters are reluctant to take this perspective, there are compelling reasons for doing so.

This opens up opportunities for using standard investment analysis techniques to rank projects competing for scarce funding. The use of these techniques lets forest managers evaluate alternative strategies and select the one which will contribute the most to the achievement of management goals. This technical note is intended to provide forest resource managers with an introduction to benefit-cost analysis and its potential use in forest management. We present a basic introduction to some important concepts of benefit-cost analysis, then discusses the nature and valuation of benefits and costs commonly found in forestry, and finally concludes with an illustrative example.

DESERTIFICATION CONTROL AND RANGELAND MANAGEMENT

The origin of the Thar desert is a controversial subject. Some consider it to be only 4000 to 10,000 years old, whereas others state that aridity started in this region much earlier. Also known as The Great Indian Desert, it is spread over four states in India, namely Punjab, Haryana, Rajasthan, and Gujarat, and two states in Pakistan and covers an area of about 4,46,000 square kilometres. The average annual rainfall of the region varies from 100 to 500 mm, it is distributed very erratically, occurring mostly between July and September. The mean average temperature varies from a minimum of 24 degrees C to 26 degrees C in summer to 4 degrees C to 10 degrees C in winter. One unique feature of this desert is that there is neither an oasis in it nor any artesian well. No native cactus or palm tree breaks the monotony of the vast expanse.

In India there are about 2.34 million km^2 of hot desert called 'Thar'. It represents one of the most inhospitable arid zones of the world, spreading mostly through western Rajasthan, Gujarat, South-Western Punjab, Haryana and part of Karnataka. About 85% of the great Indian desert lies in India and the rest in Pakistan. About 91% of the desert, *i.e.* 2.08 million km^2, falls in Rajasthan covering about 61% of the geographi-cal area of the state. The Aravali hills, older than the Himalayas, intersect the State to the north-east and in the west lies the great Indian desert the 'Thar'.

The Indian desert is characterized by high velocity wind, huge shifting and rolling sand dunes; high diurnal variation of temperature; scarce rainfall; intense solar radia-tion and high rate of evaporation. Thar desert receives between 100 to 500 mm of rainfall every year, 90% of which is received between July and September.

The sandy soils of the desert have a rapid infiltration rate of water, poor fertility, low humus con-tent due to rapid oxidation and high salinity. All conditions are very hostile for the ex-istence of life, yet, large human and livestock populations inhabit the area.

The Indian desert is highly fragile with poor primary producers but large liabilities *i.e.* the con-sumers causing severe impediments in its 'ecological regeneration' and 'desertifica-tion control' efforts. Archeological evidence suggest that the region was once a flourishing green coun-try-side with thick forests and river-systems of which 'Saraswati' and 'Yamuna' were the most important.

Epigraphic evidence by Landsat Satellites Imagery support this evidence. The onslaught of man and his domestic animals on the local ecosystem changed the panorama of the region from a land of plenty to the land of poverty in less than 5000 years. The artifacts discovered in the Pushkar and Luni basin of Rajasthan prove that man lived here as early as 5000 B.C.

Perhaps the over-exploitation of land and water resources since the earliest times has made the Rajasthan desert as a 'man-maintained' if not 'man-made'. The word desert gives the impression of a vast, tree-less undulating expanse of sand.

The great Indian desert 'Thar' does not conform to this popular notion. The desert is not an endless stretch of sand dunes, bereft of life or vegetation. During certain periods it blooms with a colourful range of trees and grasses and abounds in amazing variety of birds and animal life.

Thar desert is highly 'generic' for it become lush green with slightest precipitation. The soil is full of dor-mant seeds of various species which sprout with little moisture.

After independence, India paid much attention towards desertification control and ecological restoration of the Thar desert. A water canal called the Indira Gandhi Canal was constructed to bring the sweet Himalayan waters to the desert region. The IG Ca-nal, 649 km long, has a capacity to flow 524 m^3 of water/sec. With the canal water, the entire scenario of the Thar desert is changing fast. The 'desert ecosystem' appears to be transforming into an 'ever-green' forest ecosystem in the Command Area of the canal.

STRATEGIES TOWARDS DESERTIFICATION CONTROL IN THE THAR DESERT

Desert Afforestation Research Station to Control desertification was set up in Jodhpur in 1952. The State Forest Department made a humble beginning towards desertifica-tion control in 1958 by taking up afforestation in the Command Areas of IG canal in Ganganagar district. The major afforestation programmes implemented were: 'sand dune fixation work'; 'silvipastoral plantations'; 'village fodder and fuelwood planta-tion'; 'shelterbelt plantation'; 'ecological regeneration; restoration and rehabilitation of degraded desert lands'; 'afforestation on barren hills; re-seeding of old pastures and farm forestry'.

Introduction of Fast Growing 'Exotic' Tree Species

The indigenous tree species growing in the Thar desert are not only few in number but are also extremely slow growing. Therefore, greater attention was focused on the in-troduction and selection of fast growing exotic tree and shrub species from isoclimatic regions of the world. In this effort about 115 Eucalyptus species, 73 Acacia species and 170 miscellaneous ones from various countries including Mexico, USA, Latin America, former USSR, Africa, Israel, Peru, Kenya, Australia, Chile, Sudan and Zim-babwe and the Middle East, were introduced.

Acacia tortilis an exotic from Israel for sand dune stabilization, Prosopis juliflora, suitable for fast biomass production, Acacia nubica for sand dune stabilization, Colophospermum mopane and Dichrostachys glomerata for fodder purpose and Eucalyptus cameldulensis are few exotics suited for low rainfall area.

A number of the exotic tree species like Eucalyptus cameldulensis, E. melanophloia, Acacia tortilis, A. cillata, A. raddiana, A. senegal, A. sieberiana, A. aneura, A. salicina, Colophospermum mopane, Dichrostachys glomerata, Brasiletta millis, Schinus molis and Prosopis juliflora have emerged very promising for the In-dian desert.

Of all the exotic species tried, 'Acacia tortilis' from Israel has been ad-judged the best 'fuel-cum-fodder species' for the desert. Since its introduction, it has found a 'niche' not only in Rajasthan desert but also in other States of India. Acacia tortilis which has shown performance in growth and survival equal to or better than the indigenous Acacia senegal, is very promising. How-ever the desert dwellers are not happy. They want Acacia senegal to be planted on large scale. This species of Acacia senegal is of great socio-economic value for them. Besides yielding fodder and fuel it also gives a valuable 'gum resin'. They are also somehow linked with the chain of the great Indian majestic bird of the Thar desert called 'godawan' which is threatened with extinction. Godawan feeds on the insects which thrive on the gum resins of Acacia senegal.

STABILIZATION OF SHIFTING SAND-DUNES

In low-rainfall areas (150 mm to 400 mm), huge shifting sand dunes are commonly found, particularly near human habitations. Techniques of afforesting the shifting dunes were standardized after 10 years of experimentation.

These techniques consist of: (i) protection against biotic interferences; (ii) treatment of shifting sand-dunes by fixing barriers in parallel stripes or in a 'chess-board' design, using the local shrub material starting from the crest of the dunes to protect the seedlings from burial or ex-posure by the blowing of sand; (iii) afforestation of such treated dunes by direct seed-ling and planting. The two species commonly used for erecting 'brush-wood barriers' (micro-windbreaks) are Zizyphus nummularia and Crotalaria burhia.

The indigenous and exotic species which have proved successful in sand dune sta-bilization are: trees-Acacia senegal, Prosopis juliflora, Albizzia lebbeck, Cordia rothii, Dalbergia sissoo (in regions with mean annual rainfall of 250 mm) and Zizy-phus jujuba; shrubs-Calligonum polygonoides, Cassia auriculata, Ricinus communis and Zizyphus nummularia; grasses-Lasiurus sindicus, Panicum turgidum and Erian-thus munia. Among exotic species, Eucalyptus oleosa (Australia), Acacia tortilis (Is-rael), Parkinsonia aculeata and Acacia victoriae (Australia), Acacia albida (Middle East) were found to be very promising, especially as these species were found to be frost resistant. Acacia albida was used to stabilize 60,000 hectares of sand-dunes in the Thar desert.

Phog (Calligonum polygonoides) needs special mention. It is a very useful species of the Thar desert. It is a naturally growing shrub on the sand dunes. It has a massive network of underground root which works as effective 'sand binder'. Other species occurring on sand dunes are Aerva psuedotomentosa, Leptadenia pyrotechnica, Ci-trullus colocynthis, Lasiurus sindicus, Calotropis procera, etc. Other suitable species for planting on sand dunes are Colophospermum mopane and Prosopis cineraria.

SHELTERBELT PLANTATIONS TO REDUCE WIND VELOCITY

Shelterbelts and tree-screens consisting of a row of trees viz. Acacia tortilis, Tamarix articulata and Azadirachta indica flanked by two rows (one on each side) of smaller trees like Acacia senegal, Prosopis juliflora etc., with two rows (one on each side) of shrubs like Aerva tomentosa, Zizyphus spinachristi, Calligonum polygonoides were found to be very effective for Thar desert. In the Bikaner region rows of Eucalyptus cameldulensis and Dalbergia sissoo have also worked as effective shelterbelts and tree screens for creation of 'micro-climates'. Shelterbelts reduced the wind velocity by 20-46% on the leeward side for 2H-10H during the monsoon period (H= height of shelterbelt). Soil loss was also considerably reduced. Rajasthan State Forest Department has so far covered about 38,000 row km area under shelterbelt, road side, railway line and canal side plantations since 1978 by adopting the technol-ogy developed at CAZRI, Jodhpur.

ECOLOGICAL REGENERATION THROUGH AERIAL SEEDING

Aerial seeding of seed pellets of Cenchrus ciliaris, Acacia tortilis and Colophosper-mum mopane was done in 56 ha of military range areas in Barmer district of Rajasthan at two sites: the duny sandy plain of Jalipa and the rocky hill ranges of Jasai village. The programme was undertaken in collaboration with the military. Two methods of seeding, *i.e.*, seeding by helicopter and manual broadcasting of pellets were used. Data on establishment and growth were recorded twice, *i.e.*, at sowing and maturity stages. The manual broadcasting of pellets was found better than aerial seeding by helicopter. Under controlled conditions in the Jasai area, the plant population at sandy and rocky area were 6 and 3 individuals per m^2, respectively. The controlled sites were seeded manually.

Aerial seeding was also done in Bikaner on the left bank of the IG Canal in Sardar-pura (300 ha) and Motigarh (400 ha). The seed mixture consisted of seeds of Acacia tortilis, Colophospermum mopane, Dichrostachys nutans, Prosopis cineraria, Zizy-phus rotandifolia, Citrullus colocynthis and Lasiurus sindicus at the rate of 14 kg/ha. A. tortilis recorded highest germination and seedling density followed by C. colo-cynthis.

Also known as the Great Indian Desert, Thar Desert is located in Western India and South Eastern Pakistan. The landforms of Thar Desert is divided into three major regions - Sand Covered Thar, Plains and Hills. The Great Thar Desert spreads across the state of Rajasthan and parts of Gujarat in western India. Covering about 2,59,000sq km, the desert is interspersed with hillocks, gravel, salt marshes and some lakes.

ECOLOGICAL RESTORATION AND REGENERATION OF THE MINED WASTELANDS

Two mined wastelands, one of gypsum and the other of limestone were selected for ecological regeneration. Four plots of one hectare each were demarcated to have four treatments, *i.e.*, control, development of micro-catchment area, half moon structure development and ridge and furrow system, so that the planting pits (60 cm) might re-ceive additional rain water as run-off.

Seven indigenous and exotic species of trees and four of shrubs were selected for plantation. In general, 5 m × 5 m spacing in plant to plant and row to row was adopted. Plant species included Salvadora oleoides, S. persica, Acacia tortilis. Azadirachta indica, Prosopis juliflora, Tamarix articulata, Pithecelobium dulca, Dichrostachys nutan, Cassia stnitii, Cercidium floridun and Caesalpinnea ceraria.

The rooted slips of grasses like Cenchrus ciliaris, C. setigerus and Cymbopogon jwarancusa were also transplanted as a single row in the ridge and furrow system. 1.5 ha of mining muck heaps and the rocky substrata were used for plantation.

The muck heaps were reshaped to create slopes and inverted terraces for rainwater harvesting. At the rocky site 'half-moon' structures were developed with 3 m and 5 m spacing. More than 90% of the plants survived. Live hedges and biofences of Prosopis juliflora were raised at both the sites with 1 m × 1 m spacing to protect the plantation.

ECOLOGICAL REGENERATION THROUGH FENCING AND ENCLOSURES

The Thar desert shows tremendous resilience for regeneration when it is protected by fences and enclosures for a certain period of time. A chain of long term enclosure has been established by CAZRI and the State Forest Department on a variety of desert habitats such as hills, rocky gravelly pediments, flat pediments, sandy undulating pediments, flat aggraded older alluvial plains, sand dunes and shallow saline depres-sions. The recommended duration of protection is from 5 to 20 years. There is good growth of vegetation, both trees and grasses, upon protection. Desertification is completely halted in such regions.

POTENTIAL FOR RANGELAND DEVELOPMENT IN THE THAR DESERT

Natural regeneration of trees, shrubs and grasses in the Thar desert is very slow. The biotic pressure is so intense in the arid region of Rajasthan that the forest floor is grazed clean and the plants are browsed severely by the huge popu-lation of livestock hampering the regeneration process.

Improved Strains of Pasture Grasses and Legumes for Thar Desert

For achieving a breakthrough in the yield of different promising grasses and legumes, an exhaustive germ-plasm of over 1600 plants has been built up at the Central Arid Zone Research Institute (CAZRI) Jodhpur, Rajasthan and maintained in nursery. An active variety construction programme with manifold objective is under way. A num-ber of promising cultivars have been identified. The important cultivars are: CAZRI-358, 75, Molopo, Buffel of Cenchrus ciliaris; 76, 175, 296 of C. setigerus; 317, M-20-5, 319 of Lasiurus sindicus; 490, 491 of Dichanthium annulatum; 331, 333 of Pani-cum antidotale; 144, 1258, 1462, 1626 of Lablab purpureas; 752, 466, 1433 of Clito-ria ternatea. These varieties provide high and nutritive forage over a greater part of the year. They are drought hardy, stable, persistent and aggressive in rangeland and possess high seed yield ability, fast regeneration and good germination.

Several areas of the Thar desert are highly saline which hinders growth of grasses and palatable herbs.

Salt tolerance of various pasture species has been identified after yearly monitoring of their occurrence and yield and the results have been successfully utilized in the rangeland development in the salt affected areas of the Thar desert.

Discovery of new high yielding pasture grass varieties in the Thar desert:

1. *Cenchrus ciliaris ('Marwar Anjan'), (CAZRI-75)*: It is a selection from an entry in the germplasm received from Australia. In August 1985 CAZRI released this vari-ety for cultivation in the arid and semi-arid parts of India. Marwar Anjan is a tall, thick stemmed, erect and drought hardy perennial. Its leaves are broad, long, droopy and remain green up to maturity. It has a wide adaptability, high tillering ability and is good in regeneration. It possesses a stout root/rhizome system. It gives 2 to 3 cuttings per year. It yields 70 quintals of green fodder and 30 quintals of dry matter per hectare under desert conditions. It has 8% protein and about 60% digestibility. It produces 1 to 1.5 quintal/ha seed even after one cutting of fodder. It is persistent and aggressive variety for rangelands and remains productive for four to five years under proper management systems. It is sown at the onset of the mon-soon with the seed rate of 5 to 6 kg/ha.
2. *Cenchrus setigerus ('Marwar Dhaman'), (CAZRI-175)*: It is a selection from exotic material and well adopted in the arid and semi-arid regions of India. 'Marwar Dhaman' is excellent for grazing purpose due to its thin stem and leafy foliage. It is a drought hardy perennial grass which forms clumps at the base. It is an early ma-turing variety, flowers between 45 to 55 days, high in tillering, with good regeneration abilities and capable of giving 2 to 3 cuts per year under rainfed conditions. It provides an average yield of 40 quintal/hectare green fodder and 5 quintal/hectare dry matter in desertic regions, whereas in semi-desertic areas, the yield becomes doubled. It contains 9.5% crude protein and has 65% digestibility at half bloom stage. Its pasture remains productive for 4 to 5 years. It is moderately resistant to major insect pests..

The use of these improved grasses for reseeding the depleted 'orans' or 'village grazing lands' (gochars) assume significance for meeting the grazing needs of desert livestock.

A TRADITIONAL PRACTICE

Growing trees and grasses together has been a traditional practice in the Thar desert. It is being revived on large scale for rangeland development. Among nine tree species tried, Acacia tortilis showed survival of 98% followed by Dichanthium nutans (88%), Acacia senegal (83%) and Acacia indica (65%), whereas Prosopis cineraria and Al-bizzia lebbeck showed only 5 to 10% survival. Hardwickia binata, Colophospermum mopane and Zizyphus nummularia showed 28 to 35% survival.

Maximum height was recorded in the case of A. tortilis (291 cm) followed by A. indica (240 cm) and A. sen-egal (175 cm) and lowest in P. cineraria (77 cm). Collar diameter recorded in A. tor-tilis, A. indica and A. senegal was 8.21, 6.88 and 4.46 cm respectively.

Dry forage yield of Cenchrus setigerus in the interspaces between the trees was 2020 kg/ha in the plots of A. tortilis and 2500 kg/ha in A. indica. Dry forage yield of grass under other tree species ranged from 2750 to 2980 kg/ha as against 3000 kg/ha under pure pasture without trees.

Cenchrus ciliaris (CAZRI-75) and Cenchrus seti-gerus were sown between rows of neem (Azadirachta indica), subabool (Leucaena leucocephala) and Israeli babool (Acacia tortilis) planted in 1988 at 5 m × 5 m spacing. The growth of trees were reduced, but growth of grasses were better un-der the system. The fodder yield was more under Azadirachta indica as compared to grass under Leucaena leucocephala and Acacia tortilis.

Acacia leucophloea, A. auriculiformis, A. nilotica, A. senegal, A. tortilis, Albizzia lebbeck, Cassia siamea, Dalbergia sissoo, Derris indica, Parkinsonia aculeata, Prosopis cineraria and P. chilensis are some important multipurpose nitrogen fixing trees recommended for silvipasture and rangeland development in the Thar desert.

Under the rangeland development scheme, controlled and rotational grazing has been introduced. Some of the areas have been reseeded with perennial and nutritive grasses like Cenchrus ciliaris, Cenchrus setigerus, Dichanthium annulatum and Lasiurus sindicus.

Significant achievement has been made in the mass production of nutritive fodder 'sewan grass' (Lasiurus sindicus) for the desert livestock. This revolutionary grass, besides contributing in the development of good rangeland in the Thar desert, has sig-nificantly helped in stabilizing the blowing sand dunes and expansion of the desert. In the Jaisalmer district of Thar desert, schemes have been launched for the cultivation of green fodder in 200 ha, pasture development in 3,000 ha, and the es-tablishment of one hundred 'wood lots' and 'nurseries'.

Village Grazing Lands and Protected Fodder Lands in the Thar Desert

In the earlier days, each and every desert settlement had well guarded common prop-erty land resource such as good pasture land called 'gochar', and natural woodland called 'oran'. The 'orans' (Forests of God) were designated to honour village diety or saint and are preserved meticulously on socio-religious grounds. It is like a mini 'bio-sphere' reserve for the village and the propagules released from gochars and orans.

The basic purpose of benefit-cost analysis (BCA) is to assess the economic attractiveness of an action or project. A single project may be assessed to see whether it is profitable or meets a minimum profitability standard. BCA is also commonly used to rank alternate projects and select the most attractive option. Project assessment is based on a comparison of the costs of the project against the expected benefits. There are several ways in which the costs and benefits can be compared, but all of these measures share certain key characteristics.

In forestry projects, the benefits and costs are often incurred over a long period of time and they have to be discounted to allow for a consistent comparison. Discounting converts the value of future benefits and costs to today's dollar values. For someone with a discount rate of r percent per annum, today's value of $100 payable to that person in one year can be calculated as:

Present Value = $100/(1+r/100)

When r is ten percent, the present value equals $90.91. Looked at it in another way, $90.91 today can be seen as the present value of $100 payable one year in the future because $90.91 invested at ten percent will yield $100 in one year. Discounting is necessary because it is not accurate to treat equally dollars that are received or paid out at different times. There are three measures commonly used to compare the costs and benefits. The approach that most economists favour is simply taking the difference between the total benefits and total costs. The result is known as the present net worth (PNW) or present value of a project or action. It is the net gain of the action. The formula may be written as follows:

(1) PNW = Benefits-Costs

When one is comparing alternate projects, the project with the highest PNW is the most desirable because it provides the maximum net gain. A project with a PNW just breaks even and a negative PNW implies a net loss from the project or action.

An alternative ranking approach is to calculate the value of the benefits derived per cost dollar. This is known as the benefit/cost ratio (BCR), and may be written as:

(2) BCR = Benefits/Costs

The project with the highest BCR yields "the biggest bang for the buck" and is the most desirable. A break-even project has a BCR of one and a money-losing project has a BCR of less than one.

The third common ranking criterion is the internal rate of return (IRR), which is defined as the rate of discount that produces a PNW of zero. Again the highest IRR is associated with the best project.

Of these three approaches, the IRR will not be explored further, since it can be difficult to calculate when the investment period is very long, as is often the case in forestry. The PNW is often the best criterion for ranking projects, but the BCR is a measure that can be useful in identifying forest management alternatives that lead to higher PNWs.

Benefit-cost analysis can be used to evaluate forest management at various scales. At the smallest scale, BCA can be used to compare regeneration alternatives for a particular site. At a larger scale, alternative regeneration strategies for a management unit might be compared. For example, there is much debate over the degree to which intensive forest management should be practised and BCA can be used to compare the benefits of pursuing strategies based on different management intensities. BCA can also be used to allocate funding among management units, or competing projects. Identifying and measuring the benefits and costs is generally easier on smaller projects. The most common benefit of forest management is an increased volume of timber, either per hectare or from a management unit. Usually, the application of management changes the species that are found, especially on reforested areas, and it may lead to the production of higher quality stems. Management may also lead to increased population levels of wildlife species, increased or improved recreational opportunities, better regulation of water flow, and improve the aesthetics of an area.

Typical costs of forest management are expenses for silviculture, pest and fire protection, access and harvesting roads, and overhead. If a project is difficult to finance or can be expected to lead to cash flow problems, the cost of financing should also be included. Other costs or disbenefits may be incurred if wildlife populations or recreational opportunities are reduced, if the aesthetic quality of an area is reduced, or if water regulation is disrupted. Measuring the size of these disbenefits is often difficult but is necessary if they are to be valued. Some estimation will usually be required. Clearly, poorly planned and executed forest management activities generate costs that can be avoided.

The analyst must also select a discount rate. There are two chief options: the "real" discount rate (which has inflation netted out) and the "nominal" discount rate, which includes inflation. The advantage of excluding inflation is that there is no need to estimate future inflation levels and adjust future costs and benefits accordingly.

That leaves the analyst with the task of selecting an appropriate real interest rate. If the prime interest rate is denoted as r, the inflation rate as p, and the real interest rate as i, the relationship between these variables is given by:

(3) $(1 + r) = (1 + i) * (1 + p)$

Setting $r = 0.07$ (i.e seven percent) and $p = 0.02$ (*i.e.* two percent) results in a real rate of interest equal to 0.049 (4.9 percent). This is in the middle of the four to six percent range that is usually used for real interest rates. The valuation of these costs and benefits is more problematic. Increased timber production is perhaps the easiest benefit to value. The important point is that the benefit of a management action is the increase in timber volume, quality, or both. On Ontario management units, a low level of silviculture will not eliminate timber production altogether. Even when natural regeneration is used, there will be some timber produced. The value associated with implementing artificial regeneration is seen in the increased and more rapid production of timber, and often in the regrowth of more valuable species.

Timber can be valued at an amount equal to Crown dues, although Crown dues are not closely linked to factors that affect timber value. A more appropriate approach is to use the net value created by processing the timber. This is calculated by taking the market value of the products obtained from a cubic metre of wood and subtracting the costs of harvest, transporting the wood to the mill, processing in the mill, and transporting the product to market. A study funded by MNR's Forest Values Group used this approach to value timber in Ontario.

In districts where tendered sales are common and representative of the local timber, the bid amounts provide a basis for estimating the value of standing timber.

The analyst must be careful to ensure that the BCA has the proper scope. Stand-level analysis is appropriate for landowners who are interested in single stand management and even for larger ownerships such as conservation authorities which do not require an annual timber harvest. Furthermore, stand level results can be usefully incorporated into forest level harvest scheduling programmes when a relatively small proportion of the management unit area is affected by the treatment alternatives. However, if a manager is comparing treatments that will be applied to a significant number of hectares in a forest, then a forest level analysis is needed to properly account for between stand interactions. This type of analysis can be very complex and envolved and usually requires the use of computer modelling.

The intensification of a regeneration regime should yield a merchantable stand sooner than would otherwise be the case and presumably will contain a higher yield of a more valuable species. The benefits then will be the increased value of the future stand due to the higher stocking to the desired species and the earlier avialability of the areas for harvest.

Typically, the point at which a stand-level analysis begins is after the removal of the current stand. The rationale for this is that the money spent to regenerate the site (*i.e.* the investment) is related to objectives concerning the future stand, and it is on those merits that the regeneration expense should be evaluated. When spending more money on the harvest operation will reduce the costs of subsequent regeneration, the extra expense should be included as a regeneration cost, since this is the reason for it.

In a single stand analysis, one should ensure that there are no inconsistencies in the analysis caused by evaluating different alternatives over different time periods, as can happen when different actions will affect the stand rotation age. This can be avoided by examining a long series of repetitions of each alternative, effectively giving rise to an analysis period of infinite length. Alternatively, one could choose an analysis period based upon multiples of the harvest ages (*e.g.* compare two fifty-year harvest cycles under intensive management versus a one hundred year harvest cycle under extensive management). However, it is not always simple to find the required multiples, especially when a large number of options are being compared. This type of analysis is accomplished by considering the present net worth of a series of harvest values predicted infinetely into the future. The value of wildlife is more difficult to assess. Forest management will likely shift the distribution of wildlife species in a stand or forest. Ideally, one would like to have estimates of how wildlife populations will be affected by alternative management actions. This is difficult, due to the number of species involved and the variations that populations exhibit over time. It is not yet possible to track all species-only those of importance or value to humans are tracked (although the criteria for importance are changing rapidly).

At the present, valuation of a major species poses significant obstacles. Suppose the moose population is locally important. Moose have value because people like to view them and because people like to hunt them. In theory, it is possible to survey people who live in and/or visit the area of interest, ask what value they place on having various population levels of moose, and adjust the results to account for the entire local (human) population and tourists. Statistical refinements based on income and level of camping or hunting experience are often made.

Studies in the U.S. have shown that a typical person will be willing to pay some amount for higher populations of many species because there are better wildlife viewing opportunities and because there is some satisfaction in knowing that there are more (rather than fewer) animals of that species alive in the area. A hunter will also be willing to pay some amount to have more game species in an area because the chances of a successful hunt are increased. A larger game population will also be likely to lead to an influx of hunters from other areas, which is a plus in a regional analysis but of little effect in a provincial or national analysis.

The author is not aware of any published data based on willingness-to-pay surveys that have estimated the value of wildlife species in Ontario. Where game species are involved, one alternative is to use the value of expenditures made by hunters as a proxy for the hunting value per animal. This has the obvious drawback of ignoring the value of benefits unrelated to hunting, but since expenditure data are available, they represent the best alternative for now. Of course, the benefit-cost analysis must still relate population level to degree of hunting effort and this will require some estimation.

The valuation of recreation opportunities poses similar problems. However, the key steps are the same as in the case of wildlife and timber.

First, a quantitative relationship must be established between alternative management actions and recreation opportunities. Secondly, the value of a recreation opportunity, usually measured in terms of a recreation-day, must be estimated. One estimation technique is the willingness-to-pay survey. A more common technique, which is being employed by the MNR Forest Values Group, is known as the travel cost method. The central idea is to identify where the majority of recreation users come from, divide this region into zones of known distance to the recreation site, estimate the cost of travelling to the site from each zone, and then use the incidence of recreational use per zone and the zonal population to construct a demand curve for recreation at the site. Using site entrance fees is not appropriate for valuation purposes because prices are often set at low levels to encourage use. Finally, the valuation of such features as aesthetics is in its infancy, especially in Canada. Willingness-to-pay surveys of various types are again used in this case. Unfortunately, there are no known credible estimates of the value of aesthetic quality in Ontario.

There are two general classes of costs: out-of-pocket expenses and indirect losses or disbenefits. The out-of-pocket costs of undertaking various activities are usually unambiguous, but can be difficult to track down. Some judgement has to be exercised in the case of expenditures for fire protection and road construction and access. In these cases, a clear understanding of the reasons for making these expenditures is an important starting point.

Fire protection is intended primarily to prevent the loss of human life, and secondly to prevent the destruction of buildings, machinery, or other capital goods. Therefore, only a fraction of fire protection costs can be charged against forest management.

At a forest level the choice of the exact fraction of the fire program's costs to allocate to forest management would depend upon the fire district, proximity to habitation and recreation values, wildlife concerns and of course the value of the forest and its contribution in providing wood or forest land supplies. Any attempt at a local division of these cost allocations would be artifical due to way these costs are accounted for at the aggregate programme level, the interdependency of the social values associated with the use categories and the fact that in most areas fire prevention or control expensess are incurred to protect several use categories at one time.

Similarly, roads are built to facilitate timber extraction, subsequent assessment and regeneration, and provide access to the public for other purposes. Again, a subjective estimate of the proportion of the road value attributable to timber extraction is usually made. The costs of roads must also be amortised over some period of time, such as the 20 to 25 years often used for large capital expenditures such as pulp and paper mills.

Costs of damage or reductions in non-timber values should also be included. These can be estimated using the principles that are appropriate for estimating non-timber benefits, described above. The valuation of costs and benefits will depend on who is doing the analysis.

A planner working for a Forest Management Agreement (FMA) holder will have a different perspective than a government planner on the benefits and costs associated with alternative management strategies. For example, since some regeneration costs are partially refunded by the government on FMAs, the company and the government will naturally place a different cost on regeneration. This is as it should be. A simple example will be presented which illustrates the use of benefit-cost analysis to compare the application of three forest-level strategies on a management unit. Suppose that these strategies are characterised as:

- Reliance on Natural Regeneration
- Extensive Management
- Intensive Management

The particulars of the strategies are unimportant; the key point is that each increase in management intensity results in a greater volume of wood being produced from the unit. Suppose that the wood supply levels are expressed as annual volumes. For each option, the annual wood supply, the value of the annual harvest, and the annual cost of management. Note that wood has been valued at $\$10/m^3$, and non-timber impacts have been excluded.

It will be assumed that some harvesting will be done, and so one of the three options must be selected. The key issue is the comparison of the net gains from management intensification. The PNW and BCR associated with each option. A comparison of PNW's shows that the greatest net benefit is associated with intensive management. Therefore, if sufficient funding is available, intensive management would be preferred. However, the BCR declines as management is intensified from extensive to intensive levels. This indicates that the gain per dollar spent decreases as the management level is increased. If there are other potential investments that yield a BCR ranging from 1.46 to 1.60, then the greatest net benefit will be obtained by adopting extensive management and investing the remaining available funds in these other projects.

In this way, the BCR acts to guide investment so that the greatest overall PNW can be obtained. In this example, the profitability of management was at least partially justified by the increase of current harvest made in the expectation of future yield increases due to management. This is the allowable cut effect (ACE). Economists observe that the ACE has often been used to justify poortreatment investments. It is considered that there are really two separate decisions involved-the selection of the harvest level and the selection of the appropriate silvicultural treatment. Mainstream economists argue that these decisions should be analyzed separately on their own merits, whereas forest economists point to sustainable yield requirements as the basis for linking the two decisions.

In Ontario, the argument is clouded further by the public ownership of forests and the flow of non-timber benefits that come from them. This implies that the government should include in the analysis the social benefits that can be expected to arise from an increased harvest level. What conclusions can be drawn from this debate? The first is that one should be reluctant to include the ACE when analyzing projects on private land where the sole purpose of investing is to grow timber but ACE is a valid factor to include in analyses of public forest management.

The second point is to heed the economists' warnings and critically examine the merits of the silvicultural projects being considered. Foresters should be careful about planting on medium and poor site classes, and on sites distant from the mill. Due to the additional costs of hauling it to the mill, the value of timber grown further from the mill is lower than the value of equivalent timber close to the mill. One of the great difficulties in forest planning is dealing with the long time periods required to grow timber. The fastest growing boreal stands may be harvestable in 40 to 50 years and it is difficult to forecast real timber prices that far in the future with any accuracy. The uncertainty inherent in predictions for 100 years or more is enormous. Therefore, in view of the significant degree of uncertainty associated with assessing the profitability of a regeneration project, one may question whether such an exercise is worthwhile. Nevertheless, most people believe that planning is useful and planners can only use their best estimates. It is also the case that accurate forecasts of future prices and costs are less important when one is using BCA to select the most attractive project or to allocate funds among projects.

In these situations, any errors in forecasting will apply equally to similar projects and reservations about long-term forecasting are less relevant.

Forestry debates present society with several choices that need evaluation. The task is to estimate the value in each of the choices presented. The purpose of this chapter is to establish a framework that will allow each environmental policy option to be systematically and comprehensively assessed for its impact on the total economic welfare of the community. Such a framework consists of ensuring that the non financial values, financial values and alternative financial values derived from the forest are accurately evaluated, and hence given their full and due consideration.

There are presently significant barriers to the achievement of this objective. The three barriers identified in this report are:

1. The existing inability to determine the correct financial values for forestry operations.
2. An inability to determine the correct financial worth of alternative activities such as plantations and water production.
3. The lack of a methodology which determines the relative worth of non financial values (NFV's) of the forest resource.

Removal of these impediments, which is an exercise in microeconomic reform, will bring improvements in the financial efficiency of the Australian economy, whilst improving the overall quality of life for Australians in general. The first two impediments can be removed by the utilisation of some fairly common methods. With specific reference to the third impediment, however, there are many approaches by which appraisal of non financial values can occur. Some of these are specific methodologies such as contingent valuation and public ranking of options. Other methods centre around public involvement in the assessment of options. Yet other methods are based on using 'rules of thumb' to account for non financial values.

Among the reports surveyed, some techniques are mentioned quite frequently. These included public participation as a key necessity for getting non financial values properly assessed. Surveying of public opinion, systematic reviews of natural features and the use of experts also seemed to be relatively common. The task therefore becomes one of finding an appropriate route through the range of techniques offered.

The relationship between the concepts of externalities, economic value and community well being. The absolute requirement of good economic analysis is the need to accurately appraise all values, including externalities (non financial values) in any assessment framework. Economics provides the theoretical standard by which the existing assessment frameworks can be judged.

Several major forest reports in order to identify four significant components of forest project appraisal. These include the values of the forest, the valuation methodologies proposed for assessing values, the pricing/ financial valuation of forestry and alternative activities, and the assessment framework for appraising policy options in the search for the socially best outcome.

The role of pricing in the appraisal of forest policy options. The existing assessment processes and the various forest reports (apparently on the grounds of practicality) adopt a 'compartmentalised' approach to pricing. In other words, protection of non financial values is carried out prior to, and separate from, the pricing of forest timber. If the appraisal process is incomplete, then the pricing of timber is inadequate by definition. Pricing policy is significant for estimating value to the extent that it partly determines the financial value of forest timber.

These flaws range from a broad neglect of non financial values in project assessment to more specific issues. These include the failure to comprehensively identify and value non financial values, often even in the form of physical impacts; failure to identify or consider alternative financial values and alternative project options; and the failure to provide accurate statements about the financial viability of projects. This report makes clear that there is indeed evidence of serious weaknesses in the determination of the financial values of native forest timber. Our analysis indicates the very strong likelihood that native forest timber harvesting, as a commercial activity, is being subsidised (eg. low cost public sector loans, no commercial charges for land), is inefficient (eg.. there are better returns to be had from harvesting water in at least one case) and poorly managed (eg. the agencies are unable to provide appropriate accounting information on their commercial activities).

Consequently, a significant microeconomic reform process needs to be undertaken, if the correct financial values for native timber sales are to be determined. Inappropriate appraisal of financial values may also be suppressing economic growth in the private forestry and agroforestry sectors (and other alternative/substitute financial values). The lack of a 'level playing field', with respect to the cost of business inputs (eg. land and capital) between the public and private forest managers, suppresses investment in private forestry at a lower level than that which could normally be expected.

If financial values and alternative financial values of the forest were to be appropriately appraised, this may identify alternative revenue streams and investment opportunities for hard pressed forestry commissions. Summarised, inaccurate financial values and a limited knowledge of alternative financial values combine to undermine the basis of comparison between the financial and non financial values of the forest. This problem is compounded by the lack of a methodology for determining the relative worth of non financial values.

In conclusion, it must be reiterated that any assessment of the social benefits and costs in forest policy must include three pieces of vital data: an accurate appraisal of financial values, alternative financial values and non financial values. Without reference to such data, environmental decisions become less an attribute of good management, than of reliance on sheer blind luck.

Resolving these problems will contribute to better environmental management, greater financial performance and more jobs. This makes reform in this area a key part of the process of microeconomic change, resulting in better environmental and economic policy.

SOCIAL VALUES IN COMMERCIAL FOREST MANAGEMENT

Apart from the policy, legal and ethical reasons noted above, there are good commercial reasons to consider social values. Forest managers are increasingly subject to pressures from other interest groups, frequently concerning social values.

- Demands from forest peoples' groups may include greater respect for local populations' rights, carrying out or desisting from specific management practices, and support for their own environmental/ social projects.

- Demands from local groups to contribute to social and economic development may include support for local enterprise, employment opportunities, excision of specific areas from management, and use of company infrastructure.
- Demands from trades unions and their initiatives may include greater respect for workers' rights to organize and negotiate, fair wages and benefits, health and safety at work, and the right to skills development throughout the organization.
- Demands to recognize other forest actors' rights to monitor, control and negotiate may include the development of agreements, and procedures to manage conflicts and make compensation.

Unless an active, organized approach is taken to respond to such pressures, forest managers may find themselves facing:

- slow-downs, strikes, blockades, boycotts, legal battles;
- damage to forest stock;
- arson, sabotage or vandalism to equipment and infrastructure;
- reappropriation of forest lands for cultivation, or migration;
- development of 'cultures of resistance' among forest-dependent people and their supporters (such as some consumers).

Considerable management skill and time may need to be invested in dealing with disputes, legal challenges and compensation claims, stalled negotiations, 'bad press' and political backlashes and general hostility towards forest managers and companies.

In contrast, participatory approaches that develop people's potential contributions can benefit forest managers, by:

- improving the reputation of forest managers and companies;
- uncovering and sharing useful local information;
- broadening the base of ideas, skills and inputs applied to forestry;
- efficient apportioning of responsibility, *e.g.* local groups may be better suited to managing recreation and harvesting of non-timber forest products;
- better understanding of broader social, and thereby market, trends;
- improving transparency, accountability and therefore trust between all parties;
- longer-run cost savings and risk reduction.

By taking an active approach to people as well as to trees, forest managers greatly increase the potential social benefits from forest management and their chances of support from others. Such experience should also increase their capacity to anticipate future developments regarding social values, for example in legislation or the market, and thus to gain 'first-mover' advantage from the situation.

If some social values are indicators of market trends (and they frequently are), then it behoves forestry organizations to keep close track of them. Many companies have made good business ventures by diversifying into recreation provision in particular. And it is increasingly clear that there is considerable financial value attached to brand names of certain leading companies which are known to produce social and environmental benefits alongside fibre. For example, the Greenpeace name has been estimated to be worth hundreds of millions of dollars.

CODES OF PRACTICE AND CERTIFICATION STANDARDS ON SOCIAL ISSUES

All initiatives to define SFM principles and criteria cover social values. Some of these are becoming enshrined in legislation, while others face forest enterprises through market relations.

Current systems of forest certification specify various social requirements at the level of the forest management unit. Even so, social issues remain perhaps the most contentious of certification standards, and there are differences between standards, particularly regarding the required degree of involvement of different interest groups in forestry and treatment of peoples' rights. There are also differences in interpretation of the standards by certifiers/assessors. Whilst some confine themselves to assessing the local social outcomes of forest management, others assess the roles of local stakeholders in the management of the forest and indeed the enterprise, and call for changes that appear to reflect their own biases. Whilst it is increasingly clear that forestry should not be a principal means of social engineering, and less still should forest certification, it is generally agreed that social standards for forestry will become more stringent in future. Note that some require precise performance thresholds to be met, while others require only that the issue shall be measured. Still others consider that the social issue is a policy concern and require it to be covered in policies only.

PARTICIPATORY FOREST MANAGEMENT

With the Pace of Population booming, increased energy consumption, over exploitation of the natural resources and rapid depletion of the forest reserves accelerated natural disaster like flood, drought and cyclones in Bangladesh.

Once Bangladesh was famous for its evergreen/ semi evergreen tropical and world famous mangrove forest. But over the years due to over exploitation of forests and its non-participatory management, more than 50% of the forest resources has been depleted. Realising the grim effect of destruction of forests and to repair the lapidated environmental condition, both the government and nongovernment organization have taken up afforestation programme. The NGOs have added a new dimension in the forest management, which has ensured participation of the community people and protection of the vegetation. Although, the government has also adopted participatory forest management but due to bureaucratic attitude easy access of the poor habitants are restricted in many cases. To overcome these situations, the existing government forestry policy, which was formulated in 1994, needs radical modification. There should be room to accommodate the NGOs, grass root organisations and general people in policy formulation, execution and evaluation of the programme.

Bangladesh lies in the North-Eastern part of South Asia between 20°34' and 26°38' North in latitude and between 88°1' and 92°41' East in longitude. The total area or the country is 144,000 sq. km with a population of about 120 million, density of population is 800 person per square kilometer. The most densely populated country in the world, Bangladesh is mainly a floodplain delta, which is formed at the confluence of the Ganges, the Brahmaputra and the Meglina rivers. Natural forest represents only 6 percent of the total land area of the country and is managed and controlled by the government. Village forest which contains annual and perennial trees and still provides a major source of food and income for majority of people, is managed by private individuals.

A rapidly increasing population is placing growing demand on natural resources, especially forest sector is under pressure to become more productive and efficient to keep pace with increasing demand. At present forest in Bangladesh is in unfavourable situation in terms of meeting increasing demand and also not adequate for maintaining ecological balance. This is primarily due to heavy population pressure and limited resource base, secondly, lack of integrated planning for development of multiple resource base with active participation of people resulting in high degree of environmental degradation, as illustrated mostly by deforestation and destruction of Natural resources.

In the phase of rapid depletion of forest aggravated by increasing demand for forest resources, and considering the prevailing socio-economic condition of the country, Government has put emphasis on participatory approach in development of forest resources of the country.

FOREST SITUATION IN BANGLADESH

Bangladesh has lost over 50% of its forest resource over the period of about 25 years. Actual forest coverage is only 6 percent of the total area and the situation is worsening despite of an attempt to preserve it. At approximately 0.02 ha per person of forest, Bangladesh currently has one of the lowest per capita forest ratio in the world. In Bangladesh, government-owned forest area covers 2.19 million ha, with the remaining 0.27 million ha being privately controlled homestead forests. Of the government owned forest land, 1.49 million ha are national forests under the control of the Department of Forest, with the rest being under control of local governments. Of the state owned forests, over 90% is concentrated in 12 districts in the Eastern and South-Western region of the country. However, due to over exploitation these forests have become seriously degraded. The natural forests of the country are classified into three categories: 1) Tropical evergreen/ semi-evergreen forest in the eastern districts of Sylhet, Chittagong, Chittagong Hill Tracts, and Cox's Bazaar: 2) Moist/dry deciduous forest also known as Sal forests in the central and the northwest region and 3) Tidal mangrove forest along the coast, known as the sundarban, the largest mangrove ecosystem in the world. These forests are official reserves and placed under the jurisdiction of the Forest Department. Unfortunately, recent inventories indicate a continuing depletion of all major forests.

Forest management in Bangladesh

In Bangladesh management of government forest is the responsibility of the Forest Department under the Ministry of Environment and Forest. In this process the department is managing, protecting, developing the forest resources, forest land and also collecting the revenues. People have never been consulted nor involved in forestry activities. From the management point of view, forest of Bangladesh are being divided into three categories such as:

- State owned forest under the administrative control of Forest Department.
- State owned forest under the administrative control of Ministry of Land through District administration.
- Private village forest managed by private individuals. Forest under Forest Department control and management again divided into three major types viz; (a) Hill Forests; (b) Plain land Sal Forests, (c) Mangrove Forests.

Hill Forests: The tropical evergreen/semi evergreen forest cover as approximately 1.32 million ha of which 0.67 million ha is controlled by the forest department and rest is under the control of hill district council. Clear felling followed by replanting with suitable species (both long and short rotation) is the method of management in hill forest. Because of increased demand for timber and fuel wood and prevailing socio-economic condition of the country this forest has greatly affected and rate of denudation is considerably high. The forest department is mainly confined in raising of single species plantation. Inventory shows that most of these plantations would not give the desirable output. This programme suffers from technical, social and administrative soundness. Another problem is most of the high forest are subjected to shifting cultivation by the hill tribes. The tribes are entitled to shifting cultivation in forest land under administrative control of district administration which has resulted in the total destruction of these tropical evergreen forest. The growing stock has depleted from 23.8 million m^3 in 1964 to less than 20.7 million m^3 in 1998.

Mangrove Forests: Known as Sundarbans, the largest mangrove ecosystem in the world. Sundarban forests are being managed by selection felling method followed by natural regeneration. Beside Sundarbans, plantations are being raised with mangrove species in the newly accreted char land all along the Coast of the Bay of Bengal. Sundarban forest is an official reserve forest, unfortunately recent inventory shows a continuous depletion due to over-cutting, illegal felling. It is estimated that in less then 25 years, the volume of commercial species Sundari, Gewa, has declined by 40 to 50% respectively.

Plain land Sal Forests: Silvicultural system applied for Sal forest was coppice with standard system.

In this system matured trees were felled and the areas were protected for coppice regeneration. The typical nature of Sal forest is that this forest is scattered. In the forest areas there are agricultural lands owned by the adjacent people.

Frequently these land owners are extending their lands and encroaching to forest and in the process they are destroying the forest and subsequently converting the area to agricultural land. In this process forest lands are being marginalised day by day. FAO estimated that only 36% of the Sal forest cover remained in 1985; more recent estimates that only 10% of the forest cover remains due to over exploitation and illicit felling through there is an official base on logging since 1972. Most of the Sal forest are now substantially degraded and poorly stocked. The situation calls by for involvement of community people in the forest management.

Inventories how that there has been overall depletion in forest resources in all major state owned forest. The growing stock in Sundarban has been depleted from 20.3 million m^3 in 1960 to 10.9 million m^3 in 1998. In the Hill forest of hill districts, the growing stock has depleted from 23.8 million m^3 in 1964 to less then 20.7 million m^3 in 1998. Over-cutting by timber merchants, increased consumption linked to population growth, shifting cultivation, encroachment, illegal felling and land clearing for agriculture, lack of participatory management have been the principal causes of deforestation and shrinking of forest land in the country.

Since 1960 two major approaches regarding the role of forestry in development have been reflected in the forestry sector of Bangladesh. In the 1960's, Bangladesh as a part of Pakistan and then as an independent nation has followed 'An Industrialisation Approach' consonant with the international conventional wisdom at that time. As a result, Department of Forest raised large-scale Industrial plantation which were seen as conversion of low-yielding natural forest into artificial plantation of species (mostly teak) of great economic importance. This conversion of semi-evergreen and evergreen forest into deciduous teak plantation was largely concentrated in hill forest areas. During the plantation raising local people were not consulted and often they did not drive any benefits from these plantations. The lack of support by the local people/ communities in combination with lack of silvicultural knowledge and lack of proper maintenance contributed to raise low quality plantations and these plantations were also lost due to illegal felling. In the name of plantation the genetic resource of the ever-green/ semi-evergreen forest was lost. Forest Department was considered as revenue earning department. The main activities of Forest Department were concentrated in extraction of trees from the forest and replanting of those felled areas where applicable, Forest Department has not considered the people and their participation in managing forest of the country.

In the 1980s following a change in thinking about the role of forestry in development, and peoples participation in forestry activity was encouraged. People participation with the forestry sector realised the need of people oriented forestry programme to replenish the degraded forest resources of the country. Accordingly, in 1994 Government formulated a forest policy replacing earlier one enunciated in 1979 with a due emphasis to the need for people's participation in forest management.

PARTICIPATORY FOREST MANAGEMENT APPROACH

PAST ACTIVITIES

Forest extension activities were formally launched in the country in the year of 1964 with the establishment of two forest extension divisions at Dhaka and Rajshahi and later two divisions at Comilla and Jessore. It was really a very small programme and the activities were confined only to establish nursery in the districts headquarter and raised seedling and sell the same to individuals and organizations. The location of this programme was so urbanized and limited that it only partially served the needs of the effluent town dwellers only.

Betagi-pomora Comunity Forestry Project

The first community forestry programme in the country, started at Betagi and Pomora mouza (village) under the district of Chittagong in the year of 1979 with the personnal initiative of Prof. A. Alim, renowned forester and Prof. Dr. Mohammed Yunus, founder of Gramen Bank. Initially the project covered 160 ha of Government denuded hilly land at Betagi and with 83 landless participants from adjacent community and subsequently extend over another 205 ha of Government owned denuded hilly land at Pomora with another batch of 243 landless (families) participants. Under this programme each landless participant was provided with 1.62 ha of land for growing tree and hoticultural crops with technical and financial assistance from the Forest Department. This community programme has given the landless an identity of their own and a sense of direction in life. But this model has not been replicated in the other areas due to lack of initiative of the Forest Department as well as the Government.

Rehabilitation of Jhumia Families

Another project was undertaken by the Forest Department in the Hill tract areas to establish plantation through rehabilitation of Jhumia families in 1980. Main objectives of the programme were (i) to rehabilitate tribal families in the Unclassed State Forest(USF) lands along with rehabilitation of denuded USP land; (ii) to introduce a sustainable agroforestry production system; (iii) to improve the socio-economic condition of the tribal people and (iv) to motivate tribal people in development of forestry.

Under this programme each family was allocated 2.02 ha of USF land for growing agricultural crops (over 1.20 ha), raising plantation (0.80 ha) and for house construction (0.20 ha). The rehabilitated families were given land use rights and were allowed to enjoy 100% benefits accrued to those lands.

The participants were given input support for growing agriculture, horticulture and forestry crops and cash support for house construction. This programme continues for quite a long period of time but could not sustain mainly because of nomadic character of the tribal groups.

Another reason of failure was that the families were rehabilitated in clustered villages without considering their cultural and religious values. Thus in most of the cases it was found that the families have left the area.

A parallel programme was also initiated by the Chittagong Hill Tract Development Board in which Forest Department was responsible for implementation of afforestation component where Cittagong Hill Tract Development Board was responsible for the rehabilitation component. This programme was also not found so much responsive to hilly people except for some plantation establishment.

Development of Community Forests Project

The activities of the first phase of this project began in 1981 and were completed in 1987 in seven greater districts of the North-Western zone of the country.

The main components of the project were:

- Strip plantations along roads and highways, railways, canal sides, district and Union Parishad roads, totalling about 4,000 km.
- Fuelwood plantation on 4800 ha of depleted Government land on participatory concept.
- Agroforestry demonstration farms over 120 ha also with participatory concept.
- Replenishment of depleted homestead wood lots in 4,650 villages.
- Training of Forest Department Personnel and Village leaders.

Development of Forest Extension Services (1980-1987)

Development of Forest Extension Services (Phase II) began in 1980 with the Government funding and subsequently amalgamated in some areas (*i.e.*

North-North West district) with Asian Development Bank funded Community Forestry Project. The main activities under this programme were:

- afforestation in some 3100 villages.
- roadside tree planting along 3600 km of primary highways and roads and about 600 km of Union Parishad roads.
- production of 49 million seedlings for distribution.

Thana Afforestation and Nursery Development Project

This project is a follow-up of Development of Community Forestry Project and Forest Extension Project and has been designed primarily to: (i) increase the production of biomass fuels and (ii) enhance the institutional capability of FD and local administration in implementing a self-sustaining nationwide social forestry programme. In order to increase the production of biomass fuel and to arrest the depletion of tree resources, the project envisaged to develop tree resources base through planting of depleted sal forest as well as brining all suitable and available land in the rural areas under tree cover with active participation of the rural poor of the locality.

Originally the project was to be implemented by the Forest Department and former Thana Parishad during the period of 1987 to 1994. But in 1992 Government decided that the all project activities were to be implemented by Forest department alone. The major components of the project were:

1. Establishment of plantation over 20,225 ha depleted Sal forest areas.
2. Development of agroforestry over 4,200 ha in the Sal forest lands.
3. Raising strip plantation on 17,272 km along Road and highway, Railways, Embankment and Feeder Roads.
4. Raising 1,282 ha plantation in the land outside the BWDB.
5. Planting 7.017 million seedlings at the premises of different education, religious and social institutions
6. Establishment of 345 nurseries at Thana headquarters.
7. Raising of 10.618 million seedling for distribution to public.
8. Training of some 76,000 people of different levels.

Here this may be mentioned that at the last stage of the project implementation, the Government has found that this was quite impossible to protect the strip plantation and also impossible to trained 76,000 people by the Forest Department alone. The Government invited NGOs to participate in this programme for successful implementation. PROSHIKA, POUSH, GRAMMEN BANK and other NGOs came forward to help the Government for successful completion of the project; NGOs employed their group members to protect the strip plantation and ADAB came forward to train people at different levels with the help of its member organisations.

The above plantation activities were carried out with the direct participation of the local people with the help of the NGOs by executing benefit sharing agreement.

Coastal Greenbelt Project

Another project financed by Asian Development Bank is under implementation in the Coastal region of Bangladesh. The main objective of the project is to create a vegetative belt all along the coast to save the lives and properties of the people living in the coastal areas from devastated cyclone and tidal surges which occur very frequently in those areas. All of the activities of this project are also being carried out following participatory approach. In this project also the participants have been selected among the poor people living in the adjacent areas by involving NGO and a pre-designed benefit sharing agreements also being executed with the participants to protect their rights over plantations and to ensure benefit expected to be received out of the plantation.

Agroforestry Research Project

Pilot Agroforestry Research and Demonstration was implemented by the FD in the Sal forest areas. The project had been developed precisely to design/ develop agroforestry modules which is environmentally feasible, socio-economi-cally acceptable enhance tree and crop production at the same time to uplift the socio-economic condition of the participants. The project aimed at using 120 ha of encroached Sal Forest land of Dhaka, Mymenshing and Tangail Forest Division to develop suitable participatory plantation models.

Food Assisted Social Forestry Programme

The World Food Programme assisted the Government to develop Social Forestry as a national programme and the Government incorporated WFP assisted social forestry programme in its annual development plan from 1998. Poverty alleviation, economic rehabilitation of rural poor especially the destitute women of the society by engaging them in forestry activities, social uplift of rural poor and environmental improvement are the main objectives of this project. Historically this programme was conceived in the country since 1989 on pilot basis allocating in kind resources (Wheat) to a limited number of NGOs for raising strip plantation along roads, embankments, Highways etc. in rural areas following the participatory mechanism. In implementing this programme FD was involved later on to provide technical guidance to the NGOs and other GOB agencies. At present probably this is the largest Participatory Forestry Programme in Bangladesh. From 1990, 100 NGOs are involved in this programme and at present about 60 NGOs are continuing with the programme. Commencing from 1990 up to 1998 about 31 million trees were planted involving 0.062 million people directly and 0.62 million people indirectly. The programme has created employment to the tune of 68 million man days. This programme is being implemented by the NGOs through contractual benefit sharing among participating poor men & women 60%, NGOs 10%, the rest land owners.

PARTICIPATION IN THE PARTICIPATORY FOREST MANAGEMENT

The inclusion of communities in the management of state-owned or formerly stateowned forest resources has become increasingly common in the last 25 years. Almost all countries in Africa, and many in Asia, are promoting the participation of rural communities in the management and utilisation of natural forests and woodlands through some form of Participatory Forest Management (PFM). Many countries have now developed, or are in the process of developing, changes to national policies and legislation that institutionalise PFM.

In Bangladesh the history of NGO involvement in the field of development is not very old. After liberation, NGO started their activities through relief and rehabilitation of the war victims. During mid-seventies, NGO switched over to the socio-economic development of the rural poor, and at present there are thousands of NGOs most of whose mandate is to organise rural poor and provide awareness, education, skill training and various support services including credit to enhance participation of landless poor in the development process towards self reliance. On the basis of the networking throughout the country the NGO can be classified into two levels; (i) local and (ii) National. At present more then 100 NGOs both local and National are implementing social forestry programme in Bangladesh. It is not possible to enlist activities of all the NGOs involved in the Social forestry programme in the country. The participatory forestry activities of some of the NGOs are highlighted here who are playing pioneer role in this field.

BRAC:The Bangladesh Rural Advancement Committee (BRAC) has been established in 1972 and this is the largest NGO in Bangladesh. BRAC has six categories of projects/ programmes in broad sense among which Social Forestry falls within rural development programme. The Social/ Participatory forestry has three components; (a) Nursery establishment; (b) Plantation and (c) Establishment of agroforestry. They claim that they have assisted their women members in raising more then 225 homestead nurseries and 100 large nurseries with a combined production capacity of a million seedlings of fruits and forest seedlings. They have established 200 mulberry nurseries with the production capacity of 2 million seedlings. Social afforestation programme of BRAC is WFP assisted which has been commenced from 1989 and till now it is going on. Most of the plantation under this programme has been established along the strips. Up to this time about 33.72 million seedling have been planted over an area of 33,700 km strips along road, railway and embankment. This programme involves about 670,000 participants of which 80% are women.

PROSHIKA:Proshika-A centre for human development is one of the largest NGO in Bangladesh. The Social Forestry Programme of Proshika is a systematic intervention effort to enhance afforestation in the country and to make a case that the poor are the best managers and protectors of forest resources if they are granted usufruct rights on these resources. Proshika has introduced its group members in social forestry activities and provided them with credit and technical support, which contributed significantly to their self-sufficiency.

The main components of social forestry programme of Proshika are; i) Homestead plantation; ii) Strip and block plantation; iii) Natural Forest protection, and iv) Nursery establishment. Proshika has planted 71 million seedlings which covered along 8,887 km strips, 37,662 areas of block plantation which included natural Sal forest protection throughout the country. One of the most significant contributor of Proshika to the development areana is the introduction of the concept of participatory forest management for natural forest protection. Proshika has successfully involved the forest dwellers in the Sal forest areas of Kaliakoir, Mirzapur, Shakhipur and Shreepur thanas under district of Tangail and Gazipur for the protection of coppice Sal forest by involving group members of Proshika. It has already been proved that when poor people surviving on the forest resources are organised, trained and granted usufruct rights, they present on enormous human potential needed for afforestation and forest protection.

RDRS:The Rangpur-Dinajpur Rural Services operating in 28 thanas of greater Rangpur and Dinajpur districts covering 28 thanas. It is the largest International Integrated Rural Development NGOs operating in Northern Bangladesh for more than two decades. Its entry point in forestry was through road side plantation in 1977. Initially, seedling were protected with bamboo cages.

Situation has been changed a lot nowadays and protection of seedlings with bamboo cage has become a part of history. Besides strip plantation, they also extend their tree plantation programme in homestead, institutional grounds as well as raising of local nurseries. With the assistance of WFP they have planted about 10.66 million trees under their participatory afforestation programme.

TMSS:It stands for Thangamara Mohila Sabuj Sangha. It is an NGO exclusively meant for women. Although, it was initiated in l976, its presence was visible only since 1965. This NGO be1ieves in the concept of simple living and high thinking. TMSS was also involved in the social forestry programme particularly in the Northern districts of Bangladesh. The organisation has been implementing both road side and farm forestry with assistance from the WFP and Swiss Development Corporation (SDC) respectively.

POUSH:Another NGO has been engaged in planting in the private lands also by persuading the owner farmers in Baroibari, Kaliakoir thana with the food aid from WFP. Its activities in the forestry field are limited to strip plantations and it plans to expand its programme extensively. Target groups of POUSH are mostly landless destitute, widow and divorced women. POUSH also happens to be the first of its kind to get involved in participatory forestry in the hill district of Bandarban

PROBLEMS AND PROSPECTS OF PARTICIPATORY FOREST MANAGEMENT

Considering the demand and supply situation for forest products to meet the economic and environmental needs, no one can deny the need of people's participation in forestry. Probably there is no second answer except participatory forestry in developing, managing, and protecting the country's forest land and the forest resource. But there are numbers of issues remain unresolved. As a technical department, Forest Department is playing pioneer role in implementing and popularising Participatory Forestry in the country. Up to this time Forest Department is managed by the professional foresters who have educational background only in managing traditional forests and who do not consider people as development partners. Participatory Forestry, if we recollect the Chinese proverb, needs mental development managers towards the people.

Realisation has started among the planners, policy makers, administrators and Senior managers to involve people in forestry development activities. But up to this time Government has failed to adopt real Participatory Forestry programme to address the basic need of the peoples.

Mobilisation of the people in participatory forestry programme is another bottleneck of the Forest Department who has not had the machinery to reach the community people. NGOs who work at the grassroots level have developed their own expert to mobilise people and ensure their participation in any development programmes as partner. So NGO should be involved in the implementation of the participatory forestry programme where Forest Department should confine their activities only in technical aspect.

In the context of Bangladesh, the scarcity of land is a most vital problem. On the other hand, Forest Department controlling over 16% of the total land area of the country is still hesitant to allow Participatory Forestry in reserved forest areas. According to FD, it should be confined only in public and private lands beyond reserved forest areas through these are devoid of trees. The Participatory Forestry is being practiced in marginal lands which are under administration control of other Government departments. Recently, due to pressure from planners and donor communities, Forest Department has allowed to practice participatory forestry in Sal Forest areas. Tenure of the contract was found as a bottleneck for implementing Participatory Forestry. Forest Department allowed rights of participants over these lands for a period of seven years, but there was a strong desire, that this tenure should more and at least for rotation period, so that participants can manage and protect trees till harvesting.

A negative attitude was also observed among the Foresters to involve women in forestry activities. They viewed that activities of women should be confined in the areas where there is a locality apprehending the social problems. But in participatory forestry both men and women should be treated equally.

SUSTAINABLE FOREST MANAGEMENT INVOLVES POLITICAL AND SOCIAL PROCESSES

The past decade may come to be known by foresters as the period when they, and just about everybody else, sought to define or to prescribe sustainable forest management (SFM). Many of these efforts have painted very detailed pictures of what a well-managed forest should look like. Others have established general principles. Whatever their form, the key ingredients are sustaining multiple values through forest management, and the rights of multiple actors.

The earliest initiatives to define SFM were unilateral, undertaken by industry associations or by environmental non-governmental organizations (NGOs), and consequently were mistrusted by other stakeholders. Most of the currently accepted SFM initiatives result from multistake-holder processes, and consequently reflect a range of these stakeholders' values. They generally comprise sets of principles, criteria and indicators that have to be interpreted in detail at the local level, offering further scope for incorporating local values. They include global/intergovernmental processes (notably the UN Forest Principles 1992), regional processes (*e.g.* the Ministerial Conference on the Protection of Forests in Europe 1993) and national standards (*e.g.* the UK Forest Standards 1998). Others are led by civil society (*e.g.* the Forest Stewardship Council's Principles and Criteria). The International Institute for Environment and Development (IIED 1996) analysed 17 such initiatives and found that all had the following in common:

- sustaining yields of all socially valued goods and services from forests;
- ensuring positive impacts of forest use on different social groups;
- maintaining the state of the forest to enable continued production for future generations. These initiatives to define SFM all state or imply two premises.
  1. The purpose of forestry is to provide the highly varied needs of society; this requires that demands are signalled effectively in policy and markets, and necessitates means for making trade-offs.
  2. The basic principles of SFM need to be interpreted at the most local level at which specific goods and services are needed, *e.g.* the community level for recreation and firewood, the national level for major watersheds, and the global level for climate regulation. 'Sustain-ability' can be likened to liberty or justice, a goal we all understand and aspire to, but which has to be negotiated and defined locally to understand and achieve it in practice.

This all points to an emphasis on determining local values, on participation among interest groups, on using local decision-making processes, and on ensuring that forest management systems, markets and policies reflect social values. Only recently, however, have foresters been expected to focus on social issues. There is little prior experience of foresters taking on roles of social development, although there are historical examples: the famous *taungya* system developed in Burma in the 1850s was a response to local people's needs for farmland. Yet foresters are being held increasingly accountable for social conditions (in various new regulations, voluntary codes and certification). There are many unresolved arguments. To what extent can forestry be an instrument of social development? For which communities should foresters be held accountable for social conditions, those close to the forest or also those further away? Clearly, forest managers have at least to understand which groups are most dependent upon the forest and who could contribute most to the success or failure of forest management.

FORESTS PROVIDE MULTIPLE SOCIAL VALUES

The social values of forests differ widely, depending on the culture and social group in question and the roles which forests play in their livelihoods and quality of life. These values are not static, but change over time. Key change factors appear to be the following.

Shortages of, and opportunities to develop, the five main forms of capital (natural, physical, financial, human/individual and social/community capital). Where natural capital such as forest is abundant, it tends to be valued as a resource for conversion to generate other forms of less-abundant capital, and vice versa.

- In a similar vein, the absolute and relative scarcity of particular forest resources, such as timber or biodiversity.
- The availability and price of substitutes for forest goods and services.
- Scientific discovery, education and technology changes.
- Access rights, and resources such as labour that alter capacities to exploit forests.
- An individual's allegiance to specific groups and their shared values.
- Factors that change individual allegiances: campaigning and the political influence of certain groups, tastes and associated communications and media. The prevalence of individual consumerism has changed values away from those of the community.
- Changes in political culture.
- Changes in national development conditions.
- Changes in individual income.

Whilst the table begins with basic livelihood values such as food, and finishes with recreational and aesthetic values, it would be a mistake to assume that there is a universal, rigid hierarchy of needs. For example, for some groups, spiritual values are fundamental, while for others these are much less significant. It is therefore important to identify the relevant values from the point of view of the group in question. Indeed, certain groups define themselves (in part) by the values they hold in common. This is a source of many recent clashes between forest 'stakeholders'. These clashes can be exacerbated when some groups promote groundswells of broader opinion in favour of the values they hold most strongly, as we have witnessed with many environmental campaigning groups or forest peoples' groups. In other words, group values can be influenced by those of others. The outcome is, in large part, a function of power relations, something the forest manager may be able to do little about but must be aware of.

It is also crucial to be aware of the broader trends that affect social values. Policy, institutional and market conditions frequently make the realization of some values difficult. Furthermore, the role of forests in providing such values can be substituted by other sources (*e.g.* imported foods, non-forest employment and entertainment). The possibilities of substitution tend to be much greater for richer communities than for poorer, forest-based groups. Hence the need for an emphasis on working with the poor.

In developing countries, subsistence domestic products such as fuelwood, poles, fencing and fodder are a priority for forestry, the other priority area usually being food security from forest management or conversion: 'Over 1 billion people (about 20% of the world's population) depend either wholly or to a significant extent on forests, woodland or farm trees for their subsistence needs and/or livelihood'. As Filer and Sekhran (1998) point out, many people who are perceived as rather passive 'indigenous forest dwellers' think of themselves more as forest gardeners and forest developers, with food a number one concern. Forest managers ignore these pressing social needs at their peril.

Some demands for forest values are now far stronger than they were even a few years ago. For example, 'biodiversity' is a word that was coined only in the late 1980s, and the idea of carbon sequestration as a prime value is only beginning to be accepted. Yet now there are all sorts of international environmental agreements, especially the Biodiversity Convention and Climate Change Convention, with significant financial transfer mechanisms emerging to match the political obligations to produce biodiversity and carbon storage values. At present, those who are pressing for such 'global' forest values tend to be the politically more powerful countries and groups such as international environmental NGOs; much remains to be done to ensure that the production of these values also benefits local groups, rather than squeezing out their (more parochial) concerns.

It is quite a challenge for the forest manager to make decisions about the relative weights of the values held by different groups, from local to global, and how to integrate them or how to make choices between them where integration proves impossible. Three trends are helping here. Firstly, there is the trend for certain markets to recognize some social/environmental values of forest management that were previously unrewarded, for example the market for certified forest products and the emerging market for carbon storage. Secondly, there is development of more inclusive systems of forest policy and management decision-making, such as Round Tables, forest fora, and stakeholder liaison groups, where consultation with different groups about their values is on the increase (even if there is as yet insignificant participation in decision-making and management). Thirdly, there is the work by economists to categorize values and to develop methodologies to assess them on a single (financial) scale. While these methodologies remain contentious (it is not so much the financial magnitude that counts as who bears the cost and who gains the benefits) and indeed have so far made little difference to major policy decisions, especially those regarding private forests, the economic categorization of multiple values is useful to information science if not yet to decision science.

Broadly speaking, it includes the following.

- Direct-use values: where the value is derived directly from the forest, either in a consumptive manner (timber, nuts, fodder, game, fish, etc.) or a non-consumptive manner (tourism, recreation, etc.).
- Indirect-use values: where the value derives from environmental services such as watershed and social protection, carbon sequestration or biodiversity protection, rather than from the forest directly.
- Passive-use values: where value is accorded by the mere fact that the forest exists (existence value), or for the future possibilities that the forest represents (option value) or for the ability to pass the forest on to future generations (bequest values).

THE CHALLENGE OF ACHIEVING SECURITY OF FOREST VALUES

Despite the vast range of values that people seek in forests, few values have been evident in policies, markets and institutions until recently. This is largely because of the prevalence of governments and corporations in power structures and their relatively narrow requirements from forests (notably timber and foreign exchange).

Over time, the procedures of forest authorities that serve these interests have often become increasingly anomalous in the face of needs for multiple values. Their procedures have tended to become objectives in themselves, officers' rewards being based on observing hierarchical norms of behaviour and not on innovation or results, with an emphasis on narrow technical issues leading to a legitimation of increasing isolation from other forest interest groups and their values. The continued allocation and management of forests for timber alone, bolstered by such 'fortress forestry' authorities, hinders stakeholders from using forests to produce other goods and services.

There are usually several overlapping customary and legal interests attached to a piece of forest territory. Rarely is more than one set of interests, usually timber or mining, taken into consideration in forest policy, planning and land allocation. This is often exacerbated by local stakeholder rights having been removed through legislation. Moreover, because many governments tend to be weak, powerful (private sector) operators can appropriate these same rights. The net result is inequity in bearing the costs of forest use, and in enjoying the benefits. Means for dialogue, negotiation and partnerships between stakeholders that might improve the situation are weak; thus either the *status quo* is perpetuated or clashes over forest values escalate.

Consequently, many groups are experiencing insecurities in supplies of forest goods and services. These insecurities will worsen as population grows, or more particularly as the consumption of richer groups increases, whilst forest resources continue to diminish. By the middle of this century, if present trends continue, there will be twice as many people relying on a much smaller area of forest. It is imperative that demands for wood production in particular are consistent with supplying the other forest values that people need, especially those that currently have no market or can never operate within a market, such as local livelihood needs.

In agriculture, the concept of 'food security' at household, national and global levels helped to galvanize action against food shortages and famine.

The concept of 'security of forest goods and services' may help in making more precise decisions about forests, *i.e.* what goods and services forests should produce and who should share the costs and benefits. The value of the recent initiatives towards SFM principles, criteria and indicators is that they force different groups to be clear about what they seek, in terms of goods and services and the state of forest values.

The challenge, then, must be to achieve security of those forest goods and services that are socially valued at specific levels. There are significant policy and fiscal challenges regarding security of the (indirect-use) values that are required at national level, notably the production of watershed, biodiversity and other public benefits from private lands or management companies. In addition, divisions between countries of the North and South have hindered attempts to secure global forest values in the face of the assertions of many countries that forests are sovereign territory.

It is suggested that the local level, especially the community and household in forest-dependent areas, should be accorded priority, for forests are key for livelihood security and poverty reduction; furthermore, local forest producers are often best placed to generate the values that are required at national and global levels (although they will require incentives).

Forestry that improves social values and reduces poverty will be of a type that empowers poorer groups as forest stewards, improving their security of access to forest goods and services necessary for livelihoods. It will also mitigate or restrain the power of those in whose hands forests are currently over-concentrated and who are wasting or asset-stripping them.

PEOPLE'S MEANS TO ACHIEVE SECURITY OF FOREST VALUES

People have always planted and managed trees *where* they want, for the goods and services they want, *if* they have the means at their disposal, *i.e.* the technology and resources, and recognized rights, responsibilities, rewards and relationships with other groups.

Traditional knowledge

There is much discussion about the need to reconcile traditional indigenous knowledge with 'scientific' forestry in order to improve local action to secure forest values. It is ironic that, at a time when we are seeking means to sustain multiple values, many traditional management systems that do precisely this are on the decline. For example, Lacandones forest farmers in Mexico cultivate 75 species in a single hectare, and a farmer will need to clear no more than 10 ha of rain forest in his entire farming career. In Sweden, until the nineteenth century, rotational shifting cultivation sustained grain, root crops and livestock as well as timber, and the government saw it as a legitimate form of forest management. As Ghillean Prance has said, traditional forest peoples 'who depend upon so many of the species in the forest will hardly want to destroy more than the absolute minimum necessary for their agriculture. The settler from outside the region, on the other hand, is perplexed by such diversity and tends to destroy rather than use forest [diversity]'

Traditional forest management systems comprise both knowledge and rules, some of which may be transferable to new SFM approaches. The challenge is partly a democratic one: to admit local groups into policy and management decisions and to ensure that their rights can be exercised. But it is also a challenge to the prevailing systems of knowledge, which by and large are scientific and commercial, to admit such knowledge into the pantheon of accepted practices that guides our forest management.

Rights

Different groups have particular rights associated with obtaining forest values. Some rights directly concern the forest and its use:

- territorial rights;
- ownership of trees and other resources, such as minerals, grazing and wildlife;
- rights of access, use and control (for specific purposes and times);
- intellectual property rights.

Several problems are commonly associated with rights. One is that different rights overlap on one territory. Furthermore, many rights may not be established in legal documents, especially customary rights.

Other relevant stakeholder rights concern communities and their cohesion:

- rights to protection of cultural heritage, landscapes and folklore;

- collective rights, *e.g.* a community's rights to represent itself through its own institutions;
- rights to religious freedom;
- rights to self-determination and to development;
- rights to privacy.

National laws tend to cover specific social issues linked to forests, notably rights to public consultation (planning laws), rights of access (public rights of way) and forest worker rights, etc. Internationally, there is a growing body of relevant legally binding agreements, to which countries may be party, and which eventually find their way into national legislation. These include:

- UN Covenant on Economic, Social and Cultural Rights (1966);
- UN Covenant on Civil and Political Rights (1966);
- UN Declaration on the Human Right to Development (1986);
- International Labour Organization Convention 169 concerning indigenous and tribal peoples in independent countries (1989);
- UN Convention on Biological Diversity (1992). Many new laws are based on recent legal principles deriving from international consensus on sustainable development (such as the 'intergener-ational equity', 'polluter pays' and 'precautionary' principles). Laws on land, environment and forestry are being reviewed in many countries, especially regarding the links or conflicts between formal and customary rights.

Some groups are forceful in exerting their rights and/or have skills and resources to do so. Others are less well equipped. Irrespective of the stakeholders' powers and resources, forest managers are likely to experience problems if they ignore or violate these rights, and indeed claims. An informed approach is better. This would include anticipating future legal requirements by making provision for best practice in dealing with local groups, and continuous improvement of practice.

5

Wetland Ecosystem Conservation

TERM OF ECOSYSTEM

A system is a group of parts that interact through one or more processes. The term ecosystem was introduced and defined by Tansley (1935), who as "a fundamental organizational unit of the natural world that includes both organisms and their spatial environment." Ecosystems have since been defined in various ways, and at different spatial and temporal scales. Some ecologists define ecosystems on the basis of biotic organisms, populations, or communities. For example, Hutchinson (1978) considered the ecosystem to be the environmental context in which population or community dynamics occur. Others define ecosystems in terms of their abiotic characteristics and processes. For example, Lindeman (1942) defined ecosystems as "...the system composed of physical, chemical, and biological processes active within a space/ time unit." Regardless of whether the emphasis is on biotic components or abiotic characteristics and processes of ecosystems, both remain integral to the concept of ecosystem. Rowe (1961) emphasized this when he defined ecosystems as "...a three dimensional segment of the earth where life forms and the environment interact."

Wetland ecosystems have been defined in a variety of ways by researchers, resource managers, and regulatory authorities, depending on their specific needs and objectives. In the applied world of regulation, planning, and management, wetlands are usually defined in terms of their physical, chemical, and biological characteristics such as hydrologic regime, soil type, and plant species composition. For example, in classifying wetlands for mapping, inventory, and other purposes, Cowardin *et al.* (1979) defined wetlands as "...lands transitional between terrestrial and aquatic systems where the water table is usually at or near the surface or the land is covered by shallow water..." that are characterized by the presence of hydrophytic vegetation, hydric soils, and surface water during the growing season. Wetlands are often biodiversity 'hotspots', as well as functioning as filters for pollutants from both point and non-point sources, and being important for carbon sequestration and emissions. The value of the world's wetlands are increasingly receiving due attention as they contribute to a healthy environment in many ways. Wetland functions are defined as the normal or characteristic activities that take place in wetland ecosystems or simply the things that wetlands do. Wetlands perform a wide variety of functions in a hierarchy from simple to complex as a result of their physical, chemical, and biological attributes. For example, the reduction of nitrate to gaseous nitrogen is a relatively simple function performed by wetlands when aerobic and anaerobic conditions exist in the presence of denitrifying bacteria. Nitrogen cycling and nutrient cycling represent increasingly more complex wetland functions that involve a greater number of structural components and processes. At the highest level of this hierarchy is the maintenance of ecological integrity, the function that encompasses all of the structural components and processes in a wetland ecosystem. Wetlands are one of the most productive of all ecosystems, and carry out critical regulatory functions of hydrological processes within watersheds. Regulating water quality, water levels, flooding regimes, and nutrient and sedimentation levels are a few of these processes. As with any natural habitat, wetlands are important in supporting species diversity and have a complex of wetland values. Moreover, the pattern of seasonal variation of the wetland affects the bird population fluctuation. Even small wetlands are extremely important to the conservation of biodiversity because they provide critical breeding habitat where dispersed populations can exchange genetic material, reducing the risks of extinction.

The present review is aimed at providing in a nutshell, the distribution of wetlands, the value of Wetlands, the causes and consequences of the loss of wetlands and their conservation status with special reference to India.

DISTRIBUTION OF WETLANDS IN INDIA

In India a total area of 40494 km^2 is classified as wetlands. This consists only 1.21 per cent of the total land surface. Most of the wetlands in India are directly or indirectly linked with major river systems such as the Ganga, the Cauvery, the Krishan, the Godavari and the Tapti. A Directory of Wetlands in India (1988) gives information on the location, area and ecological categorization of wetlands of our country. Wetlands in India are distributed in different geographical regions ranging from Himalayas to Deccan plateau. The variability in climatic conditions and changing topography is responsible for significant diversity. They are classified into different types based on their origin, vegetation, nutrient status, thermal characteristics, like 1. Glaciatic Wetlands (*e.g.*, Tsomoriri in Jammu and Kashmir, Chandertal in Himachal Pradesh). Tectonic Wetlands (*e.g.*, Nilnag in Jammu and Kashmir, Khajjiar in Himachal Pradesh, and Nainital and Bhimtal in Uttaranchal). Oxbow Wetlands (*e.g.*, Dal Lake, Wular Lake in Jammu and Kashmir and Loktak Lake in Manipur and some of the wetlands in the river plains of Brahmaputra and Indo-Gangetic region. Deepor Beel in Assam, Kabar in Bihar, Surahtal in Uttar Pradesh).

The Indo-Gangetic flood plain is the largest wetland system in India, extending from the river Indus in the west to Brahmaputra in the east. This includes the wetlands of the Himalayan terai and the Indo-Gangetic plains. The vast intertidal areas, mangroves and lagoons along the 7500 kilometer long coastline in West Bengal, Orissa, Andhra Pradesh, Tamil Nadu, Kerala, Karnataka, Goa, Maharashtra and Gujarat. Mangrove forests of the Sunderbans of West Bengal and the Andaman and Nicobar Islands. Offshore coral reefs of the Gulf of Kutch, Gulf of Mannar, Lakshadweep and Andaman and Nicobar Islands. Ninety-four wetlands have been identified for conservation and management under the National Programme for Conservation and Management of Wetlands.

These wetlands are eligible for financial assistance on 100% grant basis to the concerned State Governments for undertaking activities like survey and demarcation, weed control, catchment area treatment, desiltation, conservation of biodiversity, pollution abatement, livelihood support creation of minor infrastructure, educational awareness, capacity building of various stakeholders, and community development. So far 24 States have been covered; the remaining States are expected to the covered in the Eleventh Five-Year Plan.

Wetlands play a vital role in maintaining the overall cultural, economic and ecological health of the ecosystem, their fast pace of disappearance from the landscape is of great concern. The Wildlife Protection Act protects few of the ecologically sensitive regions whereas several wetlands are becoming an easy target for anthropogenic exploitation. Survey of 147 major sites across various agro climatic zones identified the anthropogenic interference as the main cause of wetland degradation. Current spatial spread of wetlands under various categories is shown.

WETLAND LOSSES – A THREAT TO ECOLOGICAL BALANCE

Threats to wetland ecosystems comprise the increasing biotic and abiotic pressures and perils.

Biotic

(1) Uncontrolled siltation and weed infestation.

(2) Uncontrolled discharge of waste water, industrial effluents, surface run-off, etc. resulting in proliferation of aquatic weeds, which adversely affect the flora and fauna.

(3) Tree felling for fuel wood and wood products causes soil loss affecting rainfall pattern, loss of various aquatic species due to water-level fluctuation.

(4) Habitat destruction leading to loss of fish and decrease in number of migratory birds.

Abiotic

(1) Encroachment resulting in shrinkage of area.
(2) Anthropogenic pressures resulting in habitat destruction and loss of biodiversity.
(3) Uncontrolled dredging resulting in successional changes.
(4) Hydrological intervention resulting in loss of aquifers.
(5) Pollution from point and non-point sources resulting in deterioration of water quality.
(6) Ill-effects of fertilizers and insecticides used in adjoining agricultural fields.

Coastal ecosystems are among the most productive yet highly threatened systems in the world. These ecosystems produce disproportionately more services relating to human well-being than most other systems, even those covering larger total areas, but are experiencing some of the most rapid degradation and loss:

(1) About 35% of mangroves have been lost over the last two decades, driven primarily by aquaculture development, deforestation, and freshwater diversion.
(2) Some 20% of coral reefs were lost and more than a further 20% degraded in the last several decades of the twentieth century through overexploitation, destructive fishing practices, pollution and siltation and changes in storm frequency and intensity.
(3) There is established but incomplete evidence that the changes being made are increasing the likelihood of nonlinear and potentially abrupt changes in ecosystems, with important consequences for human well-being. These nonlinear changes can be large in magnitude and difficult, expensive, or impossible to reverse. For example, once a threshold of nutrient loading is crossed, changes in freshwater and coastal ecosystems can be abrupt and extensive, creating harmful algal blooms (including blooms of toxic species) and sometimes leading to the formation of oxygen-depleted zones, killing all animal life. Capabilities for predicting some nonlinear changes are improving, but on the whole scientists cannot predict the thresholds at which change will be encountered. The increased likelihood of these nonlinear changes stems from the loss of biodiversity and growing pressures from multiple direct drivers of ecosystem change. The loss of species and genetic diversity decreases the resilience of ecosystems —their ability to maintain particular ecosystem services as conditions change. In addition, growing pressures from drivers such as overharvesting, climate change, invasive species, and nutrient loading

push ecosystems toward thresholds that they might otherwise not encounter.

(4) Many wetland-dependent species in many parts of the world are in decline; the status of species dependent on inland waters and of waterbirds dependent on coastal wetlands is of particular concern. Although the evidence has geographical limitations and is chiefly from species already globally threatened with extinction.

The primary indirect drivers of degradation and loss of rivers, lakes, freshwater marshes, and other inland wetlands (including loss of species or reductions of populations in these systems) have been population growth and increasing economic development. The primary direct drivers of degradation and loss include infrastructure development, land conversion, water withdrawal, pollution, overharvesting and overexploitation, and the introduction of invasive alien species.

The current loss rates in India can lead to serious consequences, where 74% of the human population is rural and many of these people are resource dependent. Healthy wetlands are essential in India for sustainable food production and potable water availability for humans and livestock. They are also necessary for the continued existence of India's diverse populations of wildlife and plant species; a large number of endemic species are wetland dependent. Most problems pertaining to India's wetlands are related to human population. India contains 16% of the world's population, and yet constitutes only 2.42% of the earth's surface. Indian landscape has contained fewer and fewer natural wetlands over time. Restoration of these converted wetlands is quite difficult once these sites are occupied for non-wetland uses. Hence, the demand for wetland products (*e.g.*, water, fish, wood, fiber, medicinal plants etc.) will increase with increase in population. Wetland loss refers to physical loss in the spatial extent or loss in the wetland function. The loss of one km^2 of wetlands in India will have much greater impacts than the loss of one km2 of wetlands in low population areas of abundant wetlands. The wetland loss in India can be divided into two broad groups namely acute and chronic losses. The filling up of wet areas with soil constitutes acute loss whereas the gradual elimination of forest cover with subsequent erosion and sedimentation of the wetlands over many decades is termed as chronic loss.

ACUTE WETLAND LOSSES

Direct deforestation in wetlands: Mangrove vegetation are flood and salt tolerant and grow along the coasts and are valued for fish and shellfish, livestock fodder, fuel wood, building materials, local medicine, honey, bees wax and for extracting chemicals for tanning leather. Alternative farming methods and fisheries production has replaced many mangrove areas and continues to pose threats. Eighty percent of India's 4240 km^2 of mangrove forests occur in the Sunderbans and the Andaman and Nicobar Islands. But most of the coastal mangroves are under severe pressure due to the economic demand on shrimps. Important ecosystem functions such as buffer zones against storm surges, nursery grounds and escape cover for commercially important fishery are lost. The shrimp farms also caused excessive withdrawal of freshwater and increased pollution load on water like increased lime, organic wastes, pesticides, chemicals and disease causing organisms. The greatest impacts were on the people directly dependent on the mangroves for natural materials, fish proteins and revenue. The ability of wetlands to trap sediments and slow water is reduced.

Hydrologic alteration: Alteration in the hydrology can change the character, functions, values and the appearance of wetlands. The changes in hydrology include either the removal of water from wetlands or raising the land-surface elevation, such that it no longer floods. Canal dredging operations have been conducted in India from 1800s due to which 3044 km^2 of irrigated land has increased to 4550 km^2 in 1990. Initial increase in the crop productivity has given way for reduced fertility and salt accumulations in soil due to irrigated farming of arid soils. India has 32, 000 ha of peat-land remaining and drainage of these lands will lead to rapid subsidence of soil surface.

Agricultural conversion: The primary direct driver of the loss and degradation of coastal wetlands, including saltwater marshes, mangroves, seagrass meadows, and coral reefs, has been conversion to other land uses. In the Indian subcontinent due to rice culture, there has been a loss in the spatial extent of wetlands. Rice farming is a wetland dependent activity and is developed in riparian zones, river deltas and savannah areas. Due to captured precipitation for fishpond aquaculture in the catchment areas and rice-farms occupying areas that are not wetlands, water is deprived to the downstream natural wetlands. Around 1.6 million hectares of freshwater are covered by freshwater fishponds in India. Rice-fields and fishponds come under wetlands, but they rarely function like natural wetlands. Of the estimated 58.2 million hectares of wetlands in India, 40.9 million hectares are under rice cultivation.

Chronic Wetland Losses

Degradation of water quality: Water quality is directly proportional to human population and its various activities. More than 50, 000 small and large lakes are polluted to the point of being considered 'dead'. The major polluting factors are sewage, industrial pollution and agricultural runoff, which may contain pesticides, fertilizers and herbicides.

Introduced species and extinction of native biota: Wetlands in India support around 2400 species and subspecies of birds. But losses in habitat have threatened the diversity of these ecosystems. Introduction of exotic species like water hyacinth (Eichornia crassipes) and salvinia (Salvinia molesta) have threatened the wetlands and clogged the waterways competing with the native vegetation. In a recent attempt at prioritization of wetlands for conservation,

Samant (1999) noted that as many as 700 potential wetlands do not have any data to prioritize. Many of these wetlands are threatened.

Ground water depletion: Draining of wetlands has depleted the ground water recharge. Recent estimate indicates that in rural India, about 6000 villages are without a source for drinking water due to the rapid depletion of ground water.

CONDITION AND TRENDS IN WETLAND-DEPENDENT SPECIES

There is increasing evidence of a rapid and continuing widespread decline in many populations of wetland-dependent species. Data on the status and population trends of species in some inland wetland-dependent groups, including mollusks, amphibians, fish, waterbirds, and some water-dependent mammals, have been compiled and show clear declines. An overall index of the trend in vertebrate species populations has also been developed and shows a continuous and rapid decline in freshwater vertebrate populations since 1970—a markedly more drastic decline than for terrestrial or marine species.

Even in the case of more poorly known wetland fauna, such as invertebrates, existing assessments show that species in these groups are significantly threatened with extinction. For example, the IUCN Red List reports that some 275 species of freshwater crustacea and 420 freshwater mollusks are globally threatened, although no comprehensive global assessment has been made of all the species in these groups. In the United States, one of the few countries to comprehensively assess freshwater mollusks and crustaceans, 50% of known crayfish species and two thirds of freshwater mollusks are at risk of extinction, and at least one in 10 freshwater mollusks are likely to have already gone extinct. Nearly one third (1, 856 species) of the world's amphibian species are threatened with extinction, a large portion of which (964 species) are freshwater-dependent. (By comparison, just 12% of all bird species and 23% of all mammal species are threatened.)

In addition, at least 43% of all amphibian species are declining in population, indicating that the number of threatened species can be expected to rise in the future. In contrast, less than 1% of species show population increases.

Species dependent on flowing water have a much higher likelihood of being threatened than those in still water. Basins with the highest number of threatened freshwater species— between 13 and 98 species—include the Amazon, Yangtze, Niger, Paraná, Mekong, Red and Pearl (China), Krishna (India), and Balsas and Usumacinta (Central America). The rate of decline in the conservation status of freshwater amphibians is far greater than that of terrestrial species. As amphibians are excellent indicators of the quality of the overall environment, this underpins the notion of the current declining condition of freshwater habitats around the world.

Key Vulnerabilities

Gitay *et al.* (2001) have described some inland aquatic ecosystems (Arctic, sub-Arctic ombrotrophic bog communities on permafrost, depressional wetlands with small catchments, drained or otherwise converted peatlands) as most vulnerable to climate change, and have indicated the limits to adaptations due to the dependence on water availability controlled by outside factors. More recent results show vulnerability varying by geographical region. This includes significant negative impacts across 25% of Africa by 2100 (SRES B1 emissions scenario, de Wit and Stankiewicz, 2006) with both water quality and ecosystem goods and services deteriorating. Since it is generally difficult and costly to control hydrological regimes, the interdependence between catchments across national borders often leaves little scope for adaptation.

Impacts

Climate change impacts on inland aquatic ecosystems will range from the direct effects of the rise in temperature and CO2 concentration to indirect effects through alterations in the hydrology resulting from the changes in the regional or global precipitation regimes and the melting of glaciers and ice cover. Studies since the TAR (Third assessment report of IPCC) have confirmed and strengthened the earlier conclusions that rising temperature will lower water quality in lakes through a fall in hypolimnetic oxygen concentrations, release of phosphorus (P) from sediments, increased thermal stability, and altered mixing patterns. In northern latitudes, ice cover on lakes and rivers will continue to break up earlier and the ice-free periods to increase.

Higher temperatures will negatively affect micro-organisms and benthic invertebrates and the distribution of many species of fish ; invertebrates, waterfowl and tropical invasive biota are likely to shift polewards with some potential extinctions. Major changes will be likely to occur in the species composition, seasonality and production of planktonic communities (*e.g.*, increases in toxic blue-green algal blooms) and their food web interactions with consequent changes in water quality.

Enhanced UV-B radiation and increased summer precipitation will significantly increase dissolved organic carbon concentrations, altering major biogeochemical cycles. Studies along an altitudinal gradient in Sweden show that NPP can increase by an order of magnitude for a 6°C air temperature increase. However, tropical lakes may respond with a decrease in NPP and a decline in fish yields. Higher CO^2 levels will generally increase NPP in many wetlands, although in bogs and paddy fields it may also stimulate methane flux, thereby negating positive effects.

Boreal peatlands will be affected most by warming and increased winter precipitation as the species composition of both plant and animal communities will change significantly. Numerous arctic lakes will dry out with a 2-3°C temperature rise. The seasonal migration patterns and routes of many wetland species will need to change and some may be threatened with extinction. Small increases in the variability of precipitation regimes will significantly impact wetland plants and animals at different stages of their life cycle. In monsoonal regions, increased variability risks diminishing wetland biodiversity and prolonged dry periods promote terrestrialisation of wetlands as witnessed in Keoladeo National Park, India.

Wetland Management – Current Status

Wetlands are not delineated under any specific administrative jurisdiction. The primary responsibility for the management of these ecosystems is in the hands of the Ministry of Environment and Forests. Although some wetlands are protected after the formulation of the Wildlife Protection Act, the others are in grave danger of extinction. Effective coordination between the different ministries, energy, industry, fisheries revenue, agriculture, transport and water resources, is essential for the protection of these ecosystems.

CARDINAL CONSTITUENTS OF COMPREHENSIVE STRATEGY FOR WETLAND CONSERVATION

The conservation and management of wetlands calls for a comprehensive strategy, ranging from legal framework and policy support to inventorization, institutional mechanism, capacity building, and community participation. The position with regard to these aspects is as follows:

LEGAL FRAMEWORK

Though there is no separate provision for specific legal instrument for wetland conservation, the legal framework for conservation and management is provided by the following legal instruments:

1. Several legislations have been enacted which have relevance to wetland conservation. These include Forest Act, 1927, Forest (Conservation) Act, 1980, the Wildlife (Protection) Act, 1972, the Air (Prevention and Control of Pollution) Act, 1974, the Water Cess Act, 1977 and the umbrella provision of Environment (Protection) Act, 1986.
2. India has set up 505 Wildlife Sanctuaries and 100 National Parks, 14 Biosphere Reserves, 6 Heritage Sites, Projects on Tiger conservation and Elephant conservation and Marine Turtles conservation with the objective of effective conservation of wetlands, and floral and faunal wealth in forest areas.
3. Notification declaring the coastal stretches of seas, bays, estuaries, creeks, rivers and backwaters, which are influenced by tidal action (in the landward side) up to 500 metres from the high tide line, and the land between the low tide line and the high tide line as the Coastal

Regulation Zone Notification, 1991 under the provision of Environment (Protection) Act, 1986. This proposes graded restriction on setting up and expansion of industries, including pressures from human activities.

4. Portions of the listed sites have been declared as Wildlife Sanctuaries and National Parks.
5. Guidelines for sustainable development and management of brackish water aquaculture have been drawn up. State Governments like Andhra Pradesh and Tamil Nadu have aquaculture guidelines also at the local level.
6. The Biodiversity Act, 2002, and the Biodiversity Rules, 2004, are aimed at safeguarding the floral and faunal biodiversity, and regulating their flow from the country to other countries for research and commercial use. Thus, their provisions also contribute towards conserving, maintaining, and augmenting the floral, faunal and avifaunal biodiversity of the country's aquatic bodies.

Policy Support: National Environment Policy (NEP), 2006

Our National Environment Policy (NEP), approved by the Cabinet on 19 May 2006, recognizes the numerous ecological services rendered by wetlands. The NEP states:

'Wetlands are under threat from drainage and conversion for agriculture and human settlements, besides pollution. This happens because public authorities or individuals having jurisdiction over wetlands derive little revenues from them, while the alternative use may result in windfall financial gains to them.

However, in many cases, the economic values of wetlands' environmental services may significantly exceed the value from alternative use. On the otherhand, the reduction in economic value of their environmental services due to pollution, as well as the health costs of the pollution itself are not taken into account while using them as a waste dump.

There also does not yet exist a formal system of wetland regulation outside the international commitments made in respect of Ramsar sites. A holistic view of wetlands is necessary, which looks at each identified wetland in terms of its causal linkages with other natural entities, human needs, and its own attributes.'

The Environmental Policy identifies the following six-fold Action Plan:

1. Set up a legally enforceable regulatory mechanism for identified valuable wetlands to prevent their degradation and enhance their conservation. Develop a national inventory of such wetlands.
2. Formulate conservation and prudent use strategies for each significant catalogued wetland, with participation of local communities, and other relevant stakeholders.

3. Formulate and implement eco-tourism strategies for identified wetlands through multi stakeholder partnerships involving public agencies, local communities and investors.
4. Take explicit amount of impacts on wetlands of significant development projects during the environmental appraisal of such projects; in particular, the reduction in economic value of wetland environmental services should be explicitly factored into cost-benefit analysis.
5. Consider particular unique wetlands as entities with 'Incomparable Values', in developing strategies for their protection.
6. Integrate wetland conservation, including conservation of village ponds and tanks, into sectoral development plans for poverty alleviation and livelihood improvement, and the link efforts for conservation and sustainable use of wetlands with the ongoing rural infrastructure development and employment generation programmes. Promote traditional techniques and practices for conserving village ponds.

Inventorization

Survey and inventorization should take into consideration identification of different human activities, effect of both industrial and domestic effluents, and information obtained through remote sensing to be verified with the ground truth data for getting proper results.

This component includes mapping of catchment areas through revenue records, survey and assessment, and land-use pattern using GIS techniques, with emphasis on drainage pattern, vegetation cover, siltation cover, encroachment, conversion of wetlands, human settlements, total area encroached, human activities at the primary, secondary, and tertiary levels, and their impact on catchment and water body. The following surveys of wetlands have been undertaken so far:

1. Asian Wetland Directory, 1989 – identified 93 Wetlands of International Importance.
2. Wetland Directory published in 1990 by the Ministry of Environment and Forests using questionnaire survey.
3. Identification of 2167 natural freshwater wetlands covering 1.5 million ha area.
4. Identification of 65, 253 man-made freshwater wetlands covering 2.6 million ha area.
5. WWF-India and the Ministry of Environment and Forests in 1993 identified 54 additional wetlands of international importance with more details.
6. Space Application Centre using remote sensing techniques identified 27, 403 inland and coastal wetlands covering 7.6 million ha

7. Salim Ali Centre for Ornithology under UNDP project has undertaken survey of 72 districts.
8. A project on 'National Wetland Information System and Updation of Wetland Inventory' has been sanctioned by the Ministry of Environment and Forests. The objectives of this project are (1) to map and inventorize wetlands on 1:50, 000 scale by on-screen interpretation of digital IRS LISS III data of post and pre-monsoon seasons, (2) to prepare State-wise wetland Atlases, and (3) to create a digital database in GIS environment in respect of all wetlands in the country.
9. The Centre for Advanced Studies in Marine Biology at Annamalai University, Parangipettai, has been assisted in project mode for updating all wetlands in the country.

Institutional Mechanism

It is imperative to have multi-disciplinary, holistic and integrated approach for achieving long-term sustainable wetland conservation and management measures. At present, various models exist in States and different nodal agencies are responsible for implementing the Wetland Conservation Programme.

In some States, the programme is executed by the Department of Forests and/or Environment or Urban Development; in some others, it is the Department of Irrigation or Science and Technology or Fisheries. However, the Wetland Conservation and Management is a specialized technical and scientific field where multi-disciplinary approach is needed, involving a number of components like water management, sustainable fisheries development, hydrological aspects, socio-economic issues, community participation, weed control, biodiversity conservation and use of aquatic macrophytes for nutrient recycling process, hydrological aspects providing information about inflow/outflow pattern in the system, nutrient fluxes and nutritional dynamics.

These aspects need to be dealt with in a coordinated manner by managers having expertise in the relevant fields.

Taking into consideration the complexity of the issue, the State Steering Committees have been constituted under the chairmanship of Chief Secretaries of the States having members from all Departments concerned. The Committee is also expected to have representatives from communities, NGOs and academicians.

The officer from the nodal department acts as a member-secretary of the Committee. The success of the programme depends upon its strong institutional mechanism where conservation efforts are undertaken through integrated and multi-disciplinary approach. However, due to inadequacy of infrastructure and staff, conservation activities are yet to acquire comprehensiveness and sustainability in some States.

State Governments have been advised to consider constitution of Wetland Conservation Authorities so that experts from various Departments undertake conservation activities in a more scientific, cohesive and sustainable manner.

Some States have already constituted Authorities for execution of wetland conservation programmes in their respective States. Notable among them are Chilika Development Authority in Orissa (mandated to manage all identified lakes in the State); Loktak Development Authority in Manipur; Shore Area Development Authority in Andhra Pradesh; Lakes and Waterways Development Authority in Jammu and Kashmir; Lake Development Authority in Karnataka and Lake Conservation Authority in Madhya Pradesh.

Capacity Building

Capacity building is a major tool without which no conservation activity is possible. We need to have good infrastructure, trained people, and case studies to teach values and functions of wetlands in an integrated and multi-disciplinary manner. The Ministry has taken several initiatives in this regard as per details given below.

(a) It has published several reports/documents on conservation and wise use of wetlands which include six monographs on Ramsar sites in collaboration with WWF India and eco-tourism guidelines for Chilika Lake.

(b) During the Tenth Five Year Plan, several training programmes have been conducted in collaboration with different academic organizations/research institutes/State Governments/international NGOs to impart training on various components of wetland conservation which include wise use, catchment area treatment, weed control, hydrological aspects, research methodology, preparation of management action plans and community participation. Training is imparted to policy makers, senior/middle level managers, organizations, stakeholders and others. A National Training Programme for Integrated Water Resource Management and Wetland Conservation was organized during 7-11 August 2006 by Chilika Development Authority with the financial support from Ministry of Environment and Forests. More training programmes are proposed to be organized at different regions of the Country.

A series of regional workshops were organized in various parts of the country to make people aware of the importance of wetlands and integrate their traditional knowledge in the planning process. The following regional and international workshops were organized during the Tenth Plan:

1. Western Region, Gujarat
2. Southern Region, Kerala
3. Eastern Region, Orissa

4. North-Eastern Region, Manipur
5. Central Region, Madhya Pradesh
6. Northern region, Uttar Pradesh
7. Northern region, Jammu and Kashmir
8. Southern region, Lakshadweep
9. International Workshop on High Altitude Wetlands, Sikkim
10. Meeting of Board of Directors of Wetland International, Rajasthan.

Holding regional workshops along with research organizations and wetland managers is an ongoing feature.

Community Participation

(a) No decision-making is complete without participation of local people whose livelihoods depend on wetland resources. People have been using wetlands since time immemorial. We have to blend both traditional and latest scientific technologies to achieve long-term conservation goals. Participatory Rural Appraisal exercise involving local communities should be the main ingredient of community participation. It should also take into consideration issues of women and gender sensitization and involve women in the management process.

(b) The component of community participation comprises the following constituents.
   1. Assessment of resource availability by surveys and participatory rural appraisal of the site.
   2. Stakeholder analysis
   3. Contact with external institutions for resource and technical advice
   4. Utilization of wastes and aquatic weeds for energy regeneration, for example through installation of community-based biogas plants.
   5. Additional alternate income generation programmes like handloom, handicrafts, integrated farm management techniques and other measures to reduce pressure on wetlands.
   6. Highlighting of gender-related cross-cultural, governance-related practices and other special concerns for assessment by community.

(c) The Joint Forest Management Committees (JFMCs), also referred to as Village Protection Committees (VPCs) or Eco-Development Committees (EDCs), are expected to play an active role in conservation and management of wetlands located in forest fringe areas, *i.e.* normally within a radius of 5 km of forest boundary. The JFMC/VPC/EDC shall be instrumental in mobilization of communities and for implementing equitable access to information rights.

USE OF GEO-SPATIAL TECHNOLOGY IN WETLAND MANAGEMENT

Remote sensing data in combination with Geographic Information System (GIS) are effective tools for wetland conservation and management. The application encompasses water resource assessment, hydrologic modeling, flood management, reservoir capacity surveys, assessment and monitoring of the environmental impacts of water resources project and water quality mapping and monitoring.

FLOOD ZONATION MAPPING

Satellite data are used for interpretation and delineation of flood-inundated regions, flood-risk zones. Temporal data helps us to obtain correct ground information about the status of ongoing conservation projects. IRS 1C/D WIFS data having 180 km spatial resolution and high temporal repetitiveness has helped in delineating the zonation of flooding areas of large river bodies, thus helping in the preparation of state-wise and basin wise flood inventories.

Water Quality Analysis and Modeling

Remote sensing data is used for the analysis of water quality parameters and modeling. Water quality studies have been done carried out using the relationship between reflectance, suspended solid concentration, and chlorophyll-a concentration. In the near infrared wavelength range, the amount of suspended solids content is directly proportional to the reflectance. Due to spatial and temporal resolution of satellite data information of the source of pollution and the point of discharge, inflow of sewage can be regularly monitored. Using IRS LISS II data monitored the suspended load in estuarine waters of Hoogly, West Bengal in a GIS environment. In this study band 4 of the data set was found to show a wider range of digital classes indicating a better response with depth than rest of the bands. Landsat TM and IRS –1A data were used to estimate sediment load in Upper lake, Bhopal. This study showed high relationship between the satellite as well as ground truth radiometric data and total suspended solids. Different image processing algorithms are also used on Landsat MSS dataset to delineate sediment concentration in reservoirs. Qualitative remote sensing methods have been used for real time monitoring of Inland Water quality Airborne sensor has also been used to study the primary productivity and related parameters of coastal waters and large water bodies.

Water Resource Management

With the development of highly precise remote sensing techniques in spatial resolution and GIS, the modeling of watershed has become more physically based and distributed to enumerate interactive hydrological processes considering spatial heterogeneity. A distributed model with SCS curve number method called as Land Use Change (LUC) model was developed to assess the hydrological changes due to land use modification. The model developed was applied to Bagmati river catchment in Kathmandu valley basin, Nepal. The study clearly demonstrated that integration of remote sensing, GIS and spatially distributed model provides a powerful tool for assessment of the hydrological changes due to landuse modifications.

Mapping of Wetland

The Space Application Centre (SAC) has mapped the wetlands at 1:250000 scale in the mainland as well the islands using the visual interpretation of coarse resolution satellite data. The states of Sikkim, West Bengal, Goa Punjab, Haryana, Himachal Pradesh, Chandigarh, Delhi, Andaman, Nicobar, Lakshwadeep, Dadra and Nagerhaveli were mapped at 1:50000 scale. However, in the rest of the country, only wetlands of 56.25 ha and above in size could be mapped. It is known that a vast majority of wetlands-often in number, extent and conservation importance is below 50 ha in size (For example, those in the Indo-gangetic plains and in the Deccan peninsula). Thus, the inventory covered only a small number of wetlands: more over, the conservation values are not known for those wetlands even whose inventory has now been obtained. The data merely indicates location of wetlands, the classification of wetlands on 1:250, 000 scale is moreover, only geomorphologic in nature (such as Oxbow lakes, Playas, Lakes and Ponds etc.) and has no other factual biological conservation value. By itself, the information will only be partly useful for conservation of wetlands. This estimate is likely to be twice if we include wetlands of size 50 ha or less.

Threats to wetland ecosystems comprise the increasing biotic and abiotic pressures and perils. About 35% of mangroves have been lost over the last two decades, driven primarily by aquaculture development, deforestation, and freshwater diversion. Some 20% of coral reefs were lost and more than a further 20% degraded in the last several decades of the twentieth century through overexploitation, destructive fishing practices, pollution and siltation and changes in storm frequency and intensity. The primary direct driver of the loss and degradation of coastal wetlands, including saltwater marshes, mangroves, seagrass meadows, and coral reefs, has been conversion to other land uses. In the Indian subcontinent due to rice culture, there has been a loss in the spatial extent of wetlands. Wetlands in India support around 2400 species and subspecies of birds. But losses in habitat have threatened the diversity of these ecosystems Introduction of exotic species like water hyacinth (Eichornia crassipes) and salvinia (Salvinia molesta) have threatened the wetlands and clogged the waterways competing with the native vegetation. As many as 700 potential wetlands do not have any data to prioritize. Many of these wetlands are threatened. In monsoonal regions, increased variability risks diminishing wetland biodiversity and prolonged dry periods promote terrestrialisation of wetlands as witnessed in Keoladeo National Park, India. So far as current status of wetland management in India is concerned, Wetlands are not delineated under any specific administrative jurisdiction. The primary responsibility for the management of these ecosystems is in the hands of the Ministry of Environment and Forests. Although some wetlands are protected after the formulation of the Wildlife Protection Act, the others are in grave danger of extinction. Effective coordination between the different ministries, energy, industry, fisheries revenue, agriculture, transport and water resources, is essential for the protection of these ecosystems. The dynamic nature of wetlands necessitates the widespread and consistent use of satellite based remote sensors and low cost, affordable GIS tools for effective management and monitoring.

ECOLOGICAL DEGRADATION DUE TO EXPLOITATION OF NATURAL RESOURCES AND DEVELOPMENT

There is a story related to the environment, in which a boy who was a very good painter made a big painting of a scene. There were big mountains covered with trees, except a few in the far background which looked snow-covered. From one of the mountains flowed a beautiful river. Down in the valley could be seen a few cosy cottages surrounded by gardens, a few butterflies, some birds soaring high in the air and little children playing happily in the open fields. Indeed it was a beautiful painting.

The boy had a little brother. He also wanted to try his hand at painting. As soon as he got the chance, he picked up the brush and painted the mountain all black. The sparkling blue-white water of the river became brown, and with just one stroke of his brush, the flowers and trees and butterflies disappeared. His older brother was heartbroken. The only consolation was that the child who destroyed the painting was so small that he had done it in ignorance. Though it would require a lot of time and effort, the painter could paint another picture.

Something very similar is happening in our lives. How many of us have looked at nature's beauty, a thousand times more beautiful than any painting can ever be, and realised how fragile it is? Just like the child's stroke of the brush, one action of ours can destroy this beauty — be it the mountains with their forest cover or the sparkling streams and rivers of life-giving water, be it the immense sand deserts or our oceans which are teeming with life; be it our lush green rain forests or the stark beauty of our cold deserts. But, and this is an important but, there is a major difference. While the painting was spoiled by an ignorant child, our earth and its environment are being spoiled by adult humans. Even more significant is the possibility that if nature's system or what we call the environment is destroyed, the effect may be fatal for many species. Environmentalist and social activist Shri Sunderlal Bahuguna once pointed out that the agony of the present-day world is the offshoot of an illogical and indiscriminate spoliation of the sources of the earth and nature by man for his meaningless material development that is leading him fast towards destruction.

Environment: A Concept of Wholeness

The environment is a concept of wholeness (nature), with non-living and living components interdependent among themselves. It is aptly defined as 'the sum total of all conditions and influences that affect the development and life of organisms'. This comprehensive definition stresses totality, and every living organism from the lowest to the highest, including human being, has it own environment. The word 'nature' in the *Gita* also conveys the idea that it does not belong to anyone but everyone belongs to it, like a family does not belong to anyone but everyone belongs to the family. Like in a family, in the environment also interactions between its different constituents are expected, and these interactions sometimes might lead to hazardous situations. Interaction is leading to the faster deterioration of the environment.

Traditionally, our understanding of the environment was holistic. A *shloka* from the *Isha Upanishad* goes, 'the whole universe together with its creatures belongs to the Lord (nature). One can enjoy the bounties of nature by giving up all greed'. Implicit in this thought is that no creature is superior to any other, and human beings should not have absolute power over nature. Let no one species encroach on the rights and privileges of nature. The element of sustainability is ingrained in this, because the emphasis is on using nature without greed. Once the element of greed enters, exploitation starts and we cease to utilise nature for the good of all human beings.

Traditional cultures have always lived in harmony with their natural environments. Nature and humankind (*prakriti* and *purusha*) form inseparable parts of the life support system. This system has five elements: air, water, land, flora and fauna, which are interconnected, interrelated and interdependent. Deterioration in one element affects the others.

Traditional social ethics placed great emphasis on the values, beliefs and attitudes that helped man to live in harmony with nature. The *Bhumi Suktam* in the *Atharvaveda* is said to be the most impressive and eloquent testament of ecological values that can be found anywhere in world literature. These and similar texts from diverse cultural traditions throughout the earth express a world-view which is informed by the spirituality inherent in nature and stress the holistic and harmonious relationship between humanity and nature.

In the *Manusmriti* (5.45) it is written that 'he who injures innoxious beings from a wish to give himself pleasure, never finds happiness, whether living or dead'. Reference to ecological concerns is also found in *Charaka Samhita, Vimansthan*, 3.2. 'The destruction of forests is most dangerous for the nation and human beings. *Vanaspati* has a direct relation with the well-being of society. Due to the pollution of the natural environment and the destruction of forests, many diseases crop up to ruin the nation'.

During Ashoka's time (272-232 bc), perhaps for the first time in the history of the world ecological concerns became state concerns. His imperial edicts laid down rules of conduct that had to be obeyed with respect to the environment. Non-compliance was met with punishment.

T.N. Khoshoo writes, quoting Gandhiji in *Mahatma Gandhi: An apostle of applied human ecology*, that 'it is an arrogant assumption to say that human beings are lords and masters of the lower creatures. On the contrary, being endowed with greater things in life, they are the trustees of the lower animal kingdom'. The delicate and holistic balance that exists in nature has to be respected and maintained.

The Himalayas, the proverbial 'Third Pole', have always remained a source of fascination and inspiration for different people and have been deemed to be the cradle of civilisation in the subcontinent. There seems to be general agreement that the ecology of the Himalayas has been endangered. The Himalayas have exercised a great influence on the environmental conditions of northern India and the people living in the Indo-Gangetic plain. They have prevented the monsoon winds from crossing over Tibet and forced them to precipitate most of their moisture on the Indian side in the form of rain and snow. This unique ecology of the Himalayas, which has such an extensive and pervasive influence on the life of our people, needs to be preserved, conserved and qualitatively upgraded.

The developmental activities of man such as the construction of high dams, roads, exploration for minerals and mining activity and the quest for arable land have to face the challenge of intensified dynamic process, commonly referred to as geographical hazards. Natural resources are being exploited in the name of economic development. Indira Gandhi's interpretation is that the real conflict is not between environment and development but between the environment and reckless exploitation by man in the name of efficiency. We have to live a life according to the rhythm of nature. Human inference in natural environmental conditions often gives these dynamic processes catastrophic proportions, leading to disasters and irreparable damage to the natural balance of the ecosystem. It is not just concern about the extinction of the big cats, but concern for all inhabitants and non-living resources. We have to stop this undeclared war against nature. Human beings are at the crossroads. Careless application of technology is leading to eco-degradation and pollution. Gandhiji emphasised, 'The earth provides enough for every man's need but not for every man's greed'.

Sustainable development is, therefore, a concept of good and sound economic growth that can be maintained indefinitely with damage to the environment. Good environment generally begets good economics.

The words 'economics' and 'ecology' have the same root, *oikos*, which refers to a house. While economics deals with financial housekeeping, ecology deals with environmental housekeeping.

Studies have shown that the perspectives of ecology are different from those of economics in that the former stresses limits rather than continuous growth, stability rather than continuous 'development'. The ecosystem is the basic unit which has biotic and abiotic components that form an interrelated, interconnected and interdependent system. The most important characteristic of an ecosystem is that it is dynamic, evolving and auto-sustainable as long as it remains reasonably undisturbed and there is incoming sunlight. The equilibrium of an ecosystem is disturbed by external stimuli such as natural cataclysmic changes and ever-increasing human activities dictated by socio-economic growth. The basic difference is that the socio-economic system, in contrast, is hitched only to one species, human beings. In an ecosystem, different species of plants and animals including human beings and micro-organisms form an interacting system. Thus, the economic process is unidirectional and human beings can only progress forwards. Conflict between the ecosystem and the socio-economic system arises from unidirectional and unlimited human wants to meet genuine needs as also greed. This has caused ecological crisis, which in other words means human exploitation of resources at a greater rate than can be normally regenerated under natural conditions.

Central Himalayas

The central Himalayas comprise eight hill districts of Uttar Pradesh, namely Chamoli, Pauri, Tehri, Uttarkashi, Dehradoon, Almora, Nainital and Pithoragarh, spread over an area of about 52, 000 sq km. The people of the region are poor, ignorant and backward but the environment has made them simple, honest, hard-working, cheerful and courageous. The region is quite rich in religious and cultural heritage. The Hindu shrines of Badrinath, Kedarnath, Gangotri, Yamunotri and the Sikh gurudwara at Hemkund near the famous valley of flowers attracts pilgrims every year. People come not only on pilgrimage but also to escape the stresses and strains of urban life, to relax and to enjoy the beauties of nature.

Forest: A Womb

The term 'forest' applies not only to trees but also to scrub vegetation and grassland. It is aptly defined as 'a peculiar organism of unlimited kindness and extends generously the products of its life activity; it affords protection to all beings, offering shade even to the axeman, who destroys it'. Trees and forests are also important for deep psychological reasons. In returning to the forest, we are returning to the womb, not in psychoanalytical terms but in cosmological terms. We are returning to our origins. For centuries forests and the people living around have complemented each other, the latter deriving their livelihood from the farmer, who in turn maintained the ecological balance and environmental quality together with conservation of soil and water. The hill people utilise their traditional knowledge to use forest resources without destroying them. From the forest they get fuel for cooking, fodder for their cattle, fruit, timber for building their houses and medicinal herbs for curing diseases. The forest helps in maintaining the flow of perennial springs, in bringing rain, in keeping the soil and water conserved, in preventing landslides, thereby giving protection from this natural calamity. It helps regulate watershed management so as to maintain the fertility of the soil, control droughts and floods, and preserve wildlife.

Massive deforestation in the Himalayan region is the important factor in ecological degradation. Non-availability of certain species, decline of fodder and wood resources, loss of the habitat of wildlife, soil erosion, recurrent floods and drying-up springs and seasonal streams and climatic changes are the consequences of man's activity. It is obvious that there is something wrong with the management of these vital resources.

The deforestation which has taken place due to commercial exploitation of trees for timber, resin, medicinal herbs, etc., the developing of new agricultural fields, over-grazing by animals, the coming up of new habitation (*e.g.* because of the construction of the Tehri dam), the building of roads mainly after the China invasion of 1962, tourism development and other development activities, increase in the population (men as well as animals), all have had an adverse affect on the environment and have brought about ecological imbalance.

The forest has gone away from the villages. It is reported that there is a scarcity of fuel, fodder and fruit. Medicinal herbs are going to be extinct. The adverse affects noticed by us were that due to deforestation in the villages of Garhwal there is watershed failure, which has resulted in both drought and flood conditions, soil erosion, landslides, changes in the microclimate, increase in the silting rate which has caused a rise of the river beds, loss of wildlife, drying up of natural springs on which the villagers depend for drinking water.

The Chipko movement took place in April 1973 in Mandal near Gopeshwar of Chamoli district. It is a grassroots non-violent and non-political movement. It is purely an ecological movement which has brought the women of the region in the mainstream of public life, and it is guided by common rural folk and not by professional leaders. 'Chipko' means to cling to the trees to save them from being cut. It awakened among the people the need for the protection of the forests.

One aspect of the deteriorating forest ecology is the large-scale replacement of natural forests by the plantation of only commercially profitable trees. These man-made forests are not capable of working in the same way as the natural forests for maintaining the ecological balance. In some instances they may do positive harm. For example, in the Himalayan forest, the oak tree is regarded as the farmer's best friend because it absorbs water for a long time and releases it slowly. This gives rise to springs around which hill villages have been established. Its leaves are used as fodder, it has a leafy canopy and a rich undergrowth of grasses which protect the soil from being directly struck by rain, and its wood is used for making agricultural implements.

Now it is being replaced by pine trees because of their commercial use. The pine tree has not the capacity to retain water, which has resulted in the drying up of springs, creating a scarcity of drinking water. It has no canopy and no undergrowth, thus leaving the mountain slopes fully exposed to erosion by rain and wind. Its leaves are not used as fodder, and they are inflammable and acidic, which makes the land infertile. But in order to extract resin from the trees the planting of pines is going on.

WATER MANAGEMENT

With the vast majority of Kashmiri people living in villages and their income coming mainly out of land, the importance of irrigation is considerable. Being a hilly state, the problem of irrigation is complicated in many areas. At higher altitudes the main source of water is *naga* or lift irrigation. The first claim on the water of a spring is of the village near it. Paddy, the staple food, is generally grown on the fertile lands adjoining the river Vitasta, although it is produced on some higher plateaus also.

The autumn or *kharif* crops consist of maize, pulses (*moong, mash* and *motha*), etc. The spring or *rabi* crops include wheat, peas, beans and mustard. Thanks to the formation of the valley, irrigation is easy and in ordinary years abundant. If there is normal snowfall in the winter and the great mountains are well covered, the water supply for rice is sufficient. On both sides of the Vitasta, the valley rises in bold terraces, and water passes quickly from one village to another in years of good snow. In earlier times, at convenient points on the mountain, weirs or protecting snags were erected, and the water was taken into main channels which pass into small networks of ducts and eventually empty themselves in the Jhelum. Lower down in the valley, where the streams flow gently, dams were erected. On every main channel there was a *mirab* – one of the villagers – whose duty it was to maintain the system.

In earlier days, in alluvial plateaus water was scarce. Many kings of Kashmir, therefore, tried to extend irrigation facilities there. The first known work of this kind was a canal called Suvarnamani (modern Suman Kul) constructed by king Suvarna in the pre-Ashokan age. It irrigated a part of the Advin Pargana, situated on the alluvial plateau to the south of the Rembyar river.

In the eighth century a great ruler of Kashmir, King Lalitaditya, is credited with having introduced a new device called the water wheel for raising water to the higher plateaus. In this device a peasant used a long pole at the top of which was a bucket to be with water, balancing it with his feet to draw the water from a well or channel. At present this inexpensive method is used in many parts of the valley, especially for the irrigation of house gardens, and the device is now called *tol* or dip-well. With the advance of technology diesel or electric pumps are preferred for the lift irrigation of agricultural lands. For this purpose the Government of Jammu and Kashmir has a department of irrigation which looks after the lift irrigation schemes.

In Kashmir the problem of recurring floods, especially in the Vitasta, was often caused by heavy summer rains and melting of snow which flooded the arable land around. To protect cultivable lands from floods, attempts were made from early times. According to Kalhana, King Damodara built stone-lined dykes in order to guard against inundations. The minister of King Baladitya erected an embankment. The construction of embankments was meant to protect cultivable land from floods, and the surplus water thus obtained could be passed into several channels to irrigate other fields. An important attempt in this direction was made by King Lalitaditya, who arranged for distributing the waters of the Vitasta at Cakradhara (modern Tsakadar) to various villages. These drainage operations made the valley productive to some extent. But the work of irrigation started by the monarch was neglected by his incompetent successors. However, an attempt was made by Suyya, the irrigation minister of King Avantivarman in the ninth century to regulate the waters of the Vitasta and to drain the whole valley. Near Yaksadara (modern Dyargul), large rocks which had rolled down from the mountain, lining both banks, obstructed the Vitasta. Suyya dragged out the rocks and the level of the river was lowered. He thus regulated the water of the Jhelum, constructed protective works and arranged for the supply of water to each village on a permanent basis. He dammed the lake, which by its depth and well-defined boundaries was naturally designed as a great reservoir to receive the surplus waters of the dangerous floods. The endeavours of Suyya met with unique success.

Due to lawlessness and insecure conditions, agriculture steadily declined. With the advent of Islam in the valley, Hindu rule came to an end in the fourteenth century. The Muslim rulers did not give any attention to the development of the valley. Sultan Zain-ul-abidin (fifteenth century) was perhaps the only Muslim ruler who was keen to cure the miseries of the cultivators and promote their welfare. Agriculture occupied his special attention. One important measure was the construction of canals in the valley. To quote Srivara, 'There was not a piece of land, not a region and not a forest where the king did not excavate a canal. Some of these were the Kakapur canal, the Karla canal, the Chakdar canal, the Avantipur canal, the Shahkul canal (of Safapur), Lachham Kul or Zainaganga, Lall Kul or Pohri canal, Shah Kul on the Martanda canal and the Mar Canal. The Shah Kul was taken out on the left bank of the Lidder river and ran along the face of the limestone cliffs above Martanda. Here it split into four distributing channels, and finally fell over the edge of the plateau into the Jhelum valley at Anantnag. Some of these canals are still important sources of irrigation, but in most cases they have narrowed down. Before the Mar Canal was constructed, the surplus waters of the Dal Lake used to flow into the Jhelum river at Habba Kadal. This junction was closed, forcing the outflow of the lake's water into the Mar Canal, which then extended up to Shadipura. Several other earlier canals were revived and repaired, and some of them supplied water to otherwise dry Karewa lands. At present the Vitasta has narrowed, 3 which is the main cause of floods in the valley over last many years. The river needs a fresh drainage programme, and the traditional knowledge system can play an important role in its management. As tourism is an important industry in Kashmir and the Jhelum one of the main attractions, it is necessary to take up the task of drainage on a priority basis. Tourists also like to stay in the houseboats, which are either in the Jhelum or in Dal Lake.

The net result of the above irrigation projects was the draining of marshes and the reclamation of large areas for cultivation, which is essential while looking to the growth of the population.

Besides irrigation, water was used in many ways in the valley. The tradition of *pana-chaki* (*greta*) still continues in many rural areas.

We should keep in mind that the state of the environment of any place is an indication of its spiritual health, in which the deeper issues of culture, values, politics and the economic and social outlook of a community are involved. Besides, no ecosystem is altogether self-contained; it is further linked to another system and so on. The scarcity of snowfall in the winter can lead to a famine. Although nature has endowed the valley with beautiful gifts, the inner forces of matter and mind are stamping these gifts out of existence at a rapid pace.

The priorities and patterns of development, coupled with whimsical decision-making, have greatly contributed to the brutalisation of the landscape, 5 the silting of the great lakes and rivers, the pollution of air, water and soil. Not long ago, the high mountains of Kashmir supported one of the densest and richest subtropical and temperate forests of the world, covering more than 60 per cent of the total land area. But after the mid-1970s there has been a licensed massacre of green trees. Currently the jungles of Kashmir are destroyed in search of *kuth*, a highly priced medicinal plant. The smuggling of wood for building, etc., has been high for many years. As per a recent study about 91 thousand hectares of forest land were lost to various development projects during the period 1952 to 1976. The deforestation in the valley has unfortunately disturbed the ecological balance and has lessened the average snowfall. Because of it there is scarcity of irrigation and drinking water in the summer season. At many places drinking water is supplied by the Public Health Engineering Department in tanks, and irrigation is mostly dependent on rainwater. If all this is not controlled at this stage the economy will be shattered. The dream of industrialisation which we people also see in Kashmir will come to a stop, as most industries are dependent on electricity, which is produced by hydel projects in the valley. It will not be wrong to say that the economic progress of the place is largely dependent on water and its proper management.

Similarly, the problems arising from the pollution of air, water and soil caused by cement factories, stone crushers, brick kilns, smelting industries and the unchecked use of chemicals have become very serious. It seems that ecological consciousness is confined to words only. Even after 49 years of freedom there is not scientific sewerage system in the summer capital (Srinagar) of the Jammu and Kashmir state. The night-soil, which is carried on the head or in poorly maintained vehicles, spreads a foul smell all along.

The inhabitants of houseboats and the boatmen living in *dongas* dispose of their refuse in the Dal Lake or the Jhelum, thereby polluting the water. It has been shown that the drinking water supplied to some parts of Srinagar city was worse than the polluted water of the Jhelum. This polluted water does pose a serious threat, not only to human life only but also to wildlife, particularly in the Dachigam National Park.

Militancy has added fuel to the fire. It has disturbed entire ecosystem and the law of the jungle prevails these days. If this situation is not brought under control quickly the entire past of Kashmir will come to a sad end. From its seminal stages to the present state of civilisation, Kashmir has been famous for tolerance, mutual goodwill and humanism. The spirit of synthesis and assimilation is the key to Kashmiri culture. Kashmir has carved a niche for itself in the fabric of Indian culture. The present-day attempts to insulate her from humanism and synthetical modes of thinking cannot be overlooked as a mere aberration as these are aimed at destroying the essential genius of Kashmir.

6

Forests of Himalayas Conservation

CONSERVATION ISSUES IN THE HIMALAYAN REGION

The Himalayan region is bestowed with varied landscape features that provide multitude of habitats to a diverse array of faunal communities including several species of wild ungulates. The region covers nearly 11% of India's geographical area and range from sub-tropical to alpine zones. The region is well known for extensive alpine biome, temperate conifer and broadleaf forests, sub-tropical (foot-hill) forests, temperate grassy slopes, and various other habitats.

Despite tremendous potential, most of the areas in this region exhibit low abundance of wild ungulates. It is an irony that there are more Himalayan Tahr (Hemitragus jemlahicus) in New Zealand in comparison to the Himalayan region. This fact generally speaks of the status of wild ungulates in the region. Most of the ungulates are found in small isolated pockets and largely restricted to Protected Areas (PAs). Even within the PAs, especially in the eastern Himalaya and adjacent hill states, wildlife and ungulate densities are very low. It is not surprising that a naturalist, having conducted large mammal surveys in the eastern Himalaya, is known to have commented, "the stage is beautiful but many of the actors are missing".

Questions on mountain ungulate conservation in the Greater Himalaya range from 'What are the reasons for low ungulate abundance in the Himalayan region' to 'What are the major issues pertaining to their conservation' and 'What are the ways and means to achieve better conservation status for mammals in general and the ungulates in particular'. We address these issues and strategies for conservation for mountain ungulates in the Greater Himalaya in this article.

MAJOR CONSERVATION ISSUES

Several factors have led to low abundance and poor conservation status of ungulates in the Himalayan region. Firstly, the Himalayan ecosystem is relatively young, fragile and low in primary productivity. With the increase in human population the area has undergone rapid degradation, fragmentation and loss of wildlife habitat.

Poaching and trade of animal parts, competitive exclusion by the domestic livestock, faulty land use practices, and human-wildlife conflicts are other factors affecting wild ungulates and other faunal communities. Some of the issues related to conservation of wildlife ungulates.

Habitat Degradation and Fragmentation

(i) Alpine Habitats of the Greater Himalaya: The region above natural tree line (ca. 3300-3600 m above mean sea level in the western and north-western and ca. 3600-3800 m in the central and eastern Himalaya) represent alpine habitats which are characterized by treeless vegetation, alpine scrub and meadows. The ungulates frequently inhabiting this zone are Blue sheep (Pseudois nayaur), Himalayan musk deer (Moschus chrysogaster), and Himalayan tahr. The major causes leading to degradation and fragmentation of alpine habitats include overgrazing by livestock, commercial harvest of wild medicinal herbs, uncontrolled tourism and mountaineering in certain areas. The area is very prone to soil erosion, avalanches and landslides owing to steep and fragile terrain. Very few PAs in the region give complete protection to alpine habitats except a few (*e.g.*, Valley of Flowers NP and Nanda Devi National Park [NP] in Uttaranchal). However, a majority of the PAs remain neglected in terms of management even though they represent important habitats for typical faunal communities.

(ii) Sub-alpine Forests: The area between ca. 3000 m and natural 'tree line' represents an important ecological belt throughout the Himalaya. Besides Himalayan musk deer and serow (Nemorhaedus sumatraensis), it forms summer habitat for two important ungulates viz., Hangul (Cervus elaphus hanglu) in the west (Kashmir Valley) and Takin (Budorcas taxicolor) in the east (Mishmi hills, Arunachal Pradesh). Of all the habitats in the Himalayan region, the sub-alpine forests have undergone maximum degradation and fragmentation owing to anthropogenic activities such as collection of non-timber forest produce (including montane bamboo, mushroom, medicinal and aromatic plants), poaching, livestock grazing and camping by the herders (Awasthi et. al. communicated). Sathyakumar *et al.* (1993) have reported that increased livestock grazing and associated impacts have led to low musk deer densities in many areas in Kedarnath Wildlife Sanctuary. The subalpine forests, 'tree line' and the alpine scrub interspersed with alpine meadows is the optimal habitat for musk deer but this habitat, particularly the 'tree line' has degraded in many parts of the Himalaya due to cumulative impacts of livestock grazing. Most graziers prefer to camp and graze their livestock in and around 'tree line' due to availability of fuel wood, water, food and cover for livestock.

(iii) Montane forests of North-Western and Western Himalaya: The forested habitats in middle elevation rages (1500-3000 m) in the western and north-western Himalaya exhibit a great diversity of flora and fauna. The major vegetation types include Himalayan Dry Temperate (conifer), Himalayan Moist Temperate (broadleaf), and several other categories. In many areas (especially on south facing gentle slopes) the forests have been transformed into the scrub jungle and cultivation. The common ungulates of the forested habitat include Himalayan musk deer, serow, goral, sambar (Cervus unicolor), barking deer (Munitacus muntjac) and wild pig (Sus scrofa). Much of the forested habitats in the region are affected by encroachment for habitation and cultivation, livestock grazing, lopping of trees for fodder. These activities have led to failure of regeneration and resultant change in the structure and composition of forests. Rawat *et al.* (1999) have reported that plant species diversity have changed with an increase in unpalatable plant species in the fringes of Kedarnath Wildlife Sanctuary as a result of human use. The forests of Western Himalaya are rich in wild mushrooms including highly prized morel (Morchella esculenta). Local people, in large groups, visit several parts of such PAs in order to collect this mushroom thereby causing heavy disturbances and affecting the threatened and sensitive species of fauna such as Himalayan musk deer and pheasants.

(iv) Central and East Himalayan Montane Forests: Major forest types in the middle elevation (1500-3000 m) ranges of Sikkim (Central Himalaya) and Arunachal Pradesh include Temperate Broadleaf Forests, Conifer forests and bamboo brakes. These forests support Himalayan musk deer, serow, takin, several other ungulates such as Asian elephant (Elephas maximus), gaur (Bos gaurus), sambar, wild pig and various other mammals including primates, and carnivores. But for the occasional fire (localized in certain areas) and slash and burn agriculture in some parts of Arunachal Pradesh, much of the forested habitats in the central and eastern Himalaya are intact.

(v) The Shivaliks and sub-Himalayan Forests: This zone (<1500 m) represents the sub-tropical climate, varied topography, rich alluvial soils and intermingling of taxa from the Indo-malayan and Palaearctic regions. The major forest formations, according Champion & Seth (1968) include Sub-tropical Dry Evergreen Forests, Sub-tropical Pine Forests, Northern Dry Mixed Deciduous Forest, Dry Shivalik Sal

Forest, Moist Mixed Deciduous Forest, Sub-tropical Broadleaf Wet Hill Forest, Northern Tropical Semi-evergreen Forest, and Northern Tropical Wet-evergreen Forest. The Shivalik hills are best represented between the Ganges and Yamuna rivers in Uttaranchal. The entire belt all along covers an area of ca. 40,000 km^2 of which only <2100 km^2 area falls under PAs represented by Simbalbara WS, Rajaji and Corbett NPs. Ecologically, entire Shivalik belt is considered as highly sensitive zone. This region suffers heavy fragmentation and degradation of habitat due to human encroachment and proliferation of exotic weeds such as Lantana camara, Parthenium hysterophorus, Cassia tora, and Sida spp. Hajra (2002), based on the analysis of remote sensing data of Shivalik zone in Uttaranchal found that even though south facing slopes appear to be suitable for goral, but only about 13% area was moderately suitable and less than 1% area is highly suitable for this species. Rest of the area is either forested or very steep and not available for goral.

(vi) The North-Eastern Hills: The natural landscapes and habitats in this region have been extensively modified due to shifting cultivation or slash and burn (Jhum) agriculture. Besides, pressure on the land due to exploitation of forest for timber and lack of a scientific forest management has led to proliferation of exotic weeds and degradation of forests. With tremendous increase in human population during recent decades the jhum cycle has come down from 20-30 years to about 5 years and even up to 3 years in many areas. Reduction in jhum cycle due to increase in pressure on land has accelerated the process of habitat degradation and fragmentation which has affected ungulates and other faunal groups in the region.

(vii) Temperate Grassy Slopes: The temperate belt in the western and north-western Himalaya support an extensive grassland habitat which is largely anthropogenic in nature. These grassy slopes have developed largely on the south facing, steeper slopes which cannot be cultivated and are burned during winter to promote grass growth. Such slopes are grazed and maintained as hay slopes or 'Ghasnis' in many sectors of Western Himalaya. The steeper and inaccessible areas are occupied by Himalayan tahr and goral. Goral is one of the prominent species of mountain ungulates that has evolved in this habitat in the region (Mishra 1993, Pendharkar 1993). The gentler slopes close to human habitation have degraded over the years and goral habitat has been reduced considerably. The temperate grassy slopes and adjacent woodlands of Dachigam NP, Kashmir valley support a highly threatened subspecies, the Hangul or Kashmir stag. The grassy slopes of Dachigam are also reported to have degraded considerably over the years (Khursheed Ahmed, personal communication).

Habitat Loss

Excessive degradation and fragmentation eventually leads to habitat loss. Systematic studies documenting loss of ungulate habitats in the Himalayan region are lacking. Hence, it is difficult to point out clear cases of this phenomenon. However, in most of the sectors, there are plenty of evidences indicating the shrinkage of wildlife / ungulate habitat. Green (1986) reported that over 70% of potential musk deer habitat has already been lost due to habitat loss and habitat degradation in the southern side of the Greater Himalaya.

Based on the spatial time-series analysis of remote sensing data, Awasthi (2001) has pointed out that in upper regions of Bhagirathi Valley, Garhwal Himalaya there has been a considerable increase in the area of human inhabitation and cultivation during last 30 years. This study also indicates that much of the sub-alpine and temperate broadleaf forests have converted into scrub vegetation. Conversion of forested habitat into scrub, is in a way loss of habitat especially for the sensitive ungulates such as Himalayan musk deer. Displacement of human populations for the developmental projects, construction of roads along the sensitive habitats, encroachment of forests for agriculture and heavy infestation of exotic weeds are other causes of habitat loss in the region.

Competition with Domestic Livestock

Owing to high seasonality and low primary productivity, the Himalayan region supports relatively low ungulate / herbivore biomass. It is therefore, obvious that with the increase in the biomass of domestic livestock in many areas, wild ungulates have suffered competitive exclusion.

Rawat (1998) has pointed out that several areas in the Himalaya there is an overstocking of livestock leading to decreased productivity and degradation of pastures. Sathyakumar *et al.* (1993) have reported that increased livestock grazing and associated impacts have led to low musk deer densities in many areas in Kedarnath Wildlife Sanctuary. Although animal husbandry is one of the main stay of livelihood in the Himalayan region, the management of livestock especially disease surveillance, rotational grazing and pasture management have been neglected leading to conflicts with wildlife as well as PA managers.

Poaching: All mountain ungulates of the Greater Himalaya are seriously threatened due to poaching for meat, skin/hide, and for products such as the 'musk' from Musk deer.

Poaching for meat (bush meat hunting) is common in many parts of the Greater Himalaya, particularly in the Eastern Himalaya. Species such as the goral, tahr, serow, takin are poached for meat/hide by local villagers and indigenous people throughout their distribution range. There are instances of wild mountain ungulate meat served in local restaurants in towns or villages adjoining wilderness areas. Although, the extent of poaching by local villagers or indigenous people is not known, there are evidences of the consequences of high poaching levels in many areas where mountain ungulates have become either locally extinct or occur in very low densities.

Poaching for sport and meat by the personnel of the security forces in the international border areas, also have led to serious impacts on mountain ungulate populations.

Poaching of musk deer for 'musk' is rampant through out the Greater Himalaya for its high commercial value (about US $ 65,000/kg in 1985) in the international markets (Green 1986, Sathyakumar 1993b). This has led to local extinctions of this species in many parts of the Greater Himalaya. As a result, the once continuous distribution of musk deer is now confined to some isolated pockets and in most of these areas they occur in very low densities. Similarly, poaching of Himalayan tahr and goral have either resulted in local extinctions or very low densities in many areas.

GENERAL STRATEGIES FOR CONSERVATION

1. Most of the protected areas in the Himalayan region lack adequate man power and funds for proper management. In addition, several PAs have ill defined boundaries leading to conflicts between the local communities and PA management. There is an urgent need to strengthen the management of most of the PAs allocating more well trained and motivated staff, budget and infra-structure.
2. Several PAs and Reserved Forests in the Himalayan region need to be brought under community reserves where local people could be made partners in conservation and management. Through various programmes such as community based eco-tourism and participatory management of natural resources the illegal activities such as trade of animal parts and poaching could be gradually brought down.
3. There is an urgent need to increase trans-boundary co-operation between India-Nepal, India-China and India-Myanmar to control the illegal trade of wildlife products. Measures should be taken to register all the arms (licensed guns, etc) with the Wildlife warden or the Divisional Forest/Wildlife Officer.
4. Currently, there is no organised system of harvesting wild medicinal and aromatic plants and uncontrolled harvest often results in the degradation of habitat. There is a need to evolve policies related to rotational harvest of medicinal plants for the benefit of communities and there is a need to control excessive pressure on the land. As part of eco-development measure cultivation of medicinal plants needs to be promoted in the buffer zones of various PAs where human pressure for these commodities is excessive.

THE MOUNTAIN UNGULATES OF THE GREATER AND TRANS-HIMALAYA

Mountains are widely acknowledged as the 'water towers' of the world in addition to being rich repositories of, often, unique biodiversity. Recognizing these values of the mountains, the UN had declared the year 2002 as the 'International Year of the Mountains' to generate awareness about the conservation and sustainable development of the mountains.

The Himalaya are the most prominent mountains in India and the region covers ca. 12% of the country's 3.3 million km2 geographical area. The present issue is dedicated to a fascinating, yet little known part of its fauna-the mountain ungulates inhabiting the cold, rugged mountains of the higher Himalaya-the ridaks or the Mountain Monarchs, as some mountain people call them.

In this chapter, we give a brief overview of the Himalayan region, its wildlife values and the reasons for choosing the present theme for the Bulletin.

The Himalaya and Associated Mountains

The Himalaya are the youngest of world's mountain chains and have among the highest peaks in the world. These mountains form the watershed for most of the rivers flowing in northern India, which sustain millions of humans who inhabit the Indo-Gangetic plains. The high ranges of the Himalaya stop the northward flow of the monsoon clouds and thus responsible for the climate and prosperity of the people living in the northern region of the Indian subcontinent. The Himalaya trace an arc of over 2,500km, from the Nanga Parbat in the west, to the Namche Barwa in the east in a roughly NW-SE orientation. In Jammu & Kashmir, the mountains and valleys of the Pamirs and the Hindu Kush spread further west, the Karakorum, east and the Kunlun mountains towards the NE, this forming one of the most formidable mountain complexes of the world. Further north, the Altai and Tien Shan ranges lead into the heart of Cenral Asia. There are a range of low mountains further that emerge from the western fringes in Pakistan, and align in a NE-SW orientation towards the deserts of Baluchistan as the Salt and the Kirthar ranges. In the east, the mountains take a sharp southward turn from Namche Barwa into the mountains of Myanmar and Bangladesh, moving further into south east Asia. The Himalaya, are thus a part of the largest mountain complex of the world and bridges its major realms, the oriental in the east and south, the Palarctic along the north and the Ethiopian along the west.

The vast spread of the Himalaya has a width varying from 200km in parts to over 500km in others. This expanse has a great variation in topography as well as biodiversity along the south to north, and the east to west axis. Humidity in general declines from east to west and from south to north, along the Himalaya. The foothills, or the Siwalik mountains are uplifted glacial debris, at places extending to ca. 1,000m above mean sea level. Higher on are the

'Middle Himalaya' extending up to ca. 3,000m as undulating hills, at places cut steeply by flowing torrents and rivers. Beyond the Middle Himalaya, is the towering Greater Himalayan range consisting primarily of igneous formations with patches of sedimentary rocks. Bulk of this area is covered with huge glaciers and peaks, with relatively arid, cold valleys in their fold. Across this great barrier, is the vast arid expanse of the Tibetan Marginal Mountains and the Tibetan Plateau, often referred to as the Trans-Himalaya. The Trans Himalaya are categorized as the Zone 1 (with two provinces) as per the biogeographically classification by Rodgers & Panwar (1988) and roughly covers 5.6% of the country's geographical area. The rest of the Himalaya are categorized as the Zone 2 (with four provinces) and covers roughly 6.4% of the country.

More interesting facts about the Himalaya such as the orogeny of the Himalaya, ecological zonation, and flora and fauna can be found in Schaller (1977) and Polunin & Stainton (1992), apart from numerous other publications.

The Mountain Ungulates of the Himalaya and Trans-Himalaya

The ungulate fauna of the Himalaya include species such as the chital, sambar, wild pig, Asian elephant, species that are found in other parts of the country too. There are however various cervids, moschids, bovids and equuids unique to the Himalaya or limited to the high mountain chains of Central Asia. Most of these species evolved to inhabit the niches produced by spectacular mountain building during the Cretaceous and Tertiary that created new, usually cold and bleak landscapes, and are well adapted to these harsh environments. This little known fauna comprising of 15 species/subspecies have been selected for the present ENVIS Bulletin. The taxonomy of this group of animals is greatly debated and we have mostly limited our listing to classification by Schaller (1977), Schaller (1998) or Shackleton (1997). We have not included the Wild Goat (Capra aegagrus) and the Shou or Sikkim stag (Cervus elaphus wallichi), since its occurrence in the country are not confirmed.

It is noteworthy that this assemblage of species/subspecies constitutes ca. 50% of India's ungulate fauna. Many of these species are wild relatives of sheep, goat, horse/donkey and yak, thus adding value to this assemblage as an important genetic pool.

The ungulates of the high mountains are prey to charismatic predators such as the snow leopard, Tibetan wolf and common leopard. An understanding of their ecology can thus help in better management of the entire region.

Most of WII's ENVIS Bulletins so far have been taxa based (elephants, small cats, crocodilians, non-human primates) or on the PA network in the country. The issues on taxa usually provide exhaustive articles from experts with either a species or a regional focus. As alluded to earlier, in this volume we have confined ourselves to the Himalayan region among all the mountain chains in the country due to its unique value as the highest, significant ecological entity, grand and fragile, yet harbouring some of the most pristine habitats for wildlife left in the country. Within the Himalaya, we have chosen accounts on those ungulate species that are either unique to the Himalaya or are confined to the mountain ranges alone.

Layout of the Issue

The issue is divided into five Sections in order to cover various facets of mountain ungulates and their conservation. One gives species accounts of the fifteen species/subspecies present in the country. This, we believe, should be useful for the lay user as well as the serious reader to easily acquaint with the species, its distribution, status, habitat, behavioural traits, morphology and key biological facts. We have tried to give clear photographs of all species, but where not available, we have given sketches.

There is a wealth of information available with the State Forest/Wildlife Departments on animal distribution and status, their vision for wildlife conservation, and management actions being taken. Often, these efforts are little known and appreciated. The 'Protected Area Network and State Reports on Status and Management of Mountain Ungulates' is meant to bridge this gap to some extent. The Himalaya are spread over six Indian states.

The fragile Himalayan region is faced with numerous conservation issues. Some of them such as livestock grazing in Protected Areas and human-wildlife conflicts are common with the rest of the country, but there are unique socio-economic peculiarities that make dealing with them more challenging. These issues and possible solutions for them are detailed for the Himalayan and the Trans-Himalayan regions.

The Himalaya have fascinated naturalists since time immemorial. A few decades ago however some scientists and naturalists have painstakingly documented the region's wildlife, often under very harsh conditions and with few facilities. We have tried to bring this perspective on what was their driving force, their trials and tribulations in initiating these studies, in 'Semi-Scientific Accounts by Veterans on Research and Conservation Experiences on Mountain Ungulates'.

Although there have been few quantitative studies and assessments on the mountain ungulates in the Himalaya, there is a wealth of information as anecdotal accounts that reveal a lot about the species. We have developed and included an exhaustive bibliography on the useful references for mountain ungulates and about conservation of the region.

The Wildlife Institute of India has pioneered research on wildlife in the Himalayan region since its inception, the very first research project being an extensive survey on the snow leopard and its prey species in the Western Himalaya. Through the twenty years of its existence, WII researchers and scientists have worked hard under daunting conditions to conduct over 30 critical studies and 20 surveys documenting the varied facets of the region's biodiversity and conservation issues. Through this Bulletin we wish to add an important and useful compilation on the conservation of the magnificent mountains crowning the country.

CONSERVATION ISSUES IN THE TRANS-HIMALAYA

A variety of physical, biotic and political characteristics of the Trans-Himalaya influence conservation issues that are peculiar to the region. Chief among the challenges is the limited resources available to the native population of the region. These populations which mostly occur at low to moderate densities of <2 persons per km2, are primarily agro-pastoral, or, as in the Tibetan Plateau zone, are largely nomadic pastoralists. Human populations are increasing in the region with the breakup of the traditional polyandrous system and with fewer people opting for becoming celibate monks and nuns. An important factor that needs to be considered is that in the harsh Trans-Himalayan landscape there is hardly any area that is not already in use by people at some time or the other during the year. Arable land is mostly limited to alluvial fans, and some stable areas in the valley bottoms.

Almost all the available arable land is already under cultivation and all pastures are grazed by the domestic stocks at least seasonally. Addition of newer areas under some poverty alleviation schemes incorporate development of expensive and long flow irrigation systems. Our observations show that these most frequently end up as failures since most of such channels are damaged by avalanches or are washed away by floods or simply break up due to the unstable substrate. Thus in spite of numerous efforts any significant addition of arable land is remote. National parks and sanctuaries in India do not permit consumptive use and require resettlement of people outside such areas. The point that we are driving at is that the Trans-Himalayan region has very specific and limited area for cultivation or use as pastures and thus offers no or very few alternatives for resettlement of people outside PAs.

The region has other peculiarities such as very poor road access, power supply, and difficult communication, apart from a harsh climate during much of the year. These features of the region thus do not allow scope for conventional industrial development and employment in the region, as is possible in other regions of the country.

As far as the wildlife values are concerned a very important characteristic of the Trans-Himalayan area is that it provides almost continuous wildlife habitat. Almost the entire landscape has large mammals, including the snow leopard and wolf, but the densities may vary greatly from very poor areas to small pockets that may be rich in some large mammals. This means that a large amount of wildlife may actually be occurring outside existing PAs. In Nepal, for example, over 60% of the snow leopards are thought to occur outside

PAs. In India, our coarse estimate is that of the probable maximum of 600 snow leopards in India, ca. 80% may be occurring outside PAs. Among other endangered wildlife species such as the Tibetan antelope, Tibetan gazelle, Tibetan argali, brown bear, and kiang, the entire or substantial populations occur outside existing PAs. Mishra, (2001). has found that in Spiti, the Tabo area that is not within any PA has bharal densities that are higher than the Kibber Wildlife Sanctuary.

Shah (1996) has found remnant populations of the highly endangered Tibetan gazelle and argali in unprotected areas in northern Sikkim. The paradox here is that in spite of a very large proportion of the Trans Himalaya being under the PA network, numerous endangered species continue to occur outside. With this, and the fact that the region in general is resource deficient, with few livelihood options for the local herders we need to reconsider the approach of having large national parks and wildlife sanctuaries as 'inviolate areas' in the region. This is a challenging situation and the lack of a vision for conservation, as usually defined by Management Plans for PAs is absent at present.

Conservation issues common to the area primarily relate to deficiencies of infrastructure and staff for PA management, grazing competition between wild and domestic herbivores, conflicts relating to damage to crops and livestock by wildlife, some levels of poaching of snow leopard and prey species, wildlife diseases and political issues.

These issues were also flagged as the most important ones in a meeting between scientists and managers at Leh and of numerous snow leopard experts from all over the snow leopard range countries. We now examine these issues in some detail:

Infrastructure and Staff for PA Management

In the harsh environs of the region, there is a severe shortage of staff, effective infrastructure and funds for the management of the PAs. For example Ladakh has ca. 13,100 km^2 under its PA network with a total park staff of merely about 20. This translates to 655 km^2 to every park staff-a completely ineffective strength to manage the region under any circumstance, but especially so under the difficult climatic and topograhic conditions in the Trans-Himalaya.

The numerous existing tasks of the Wildlife Department range from protection, tourism management, verification of compensation claims to nature education activities in the PAs spread all over the ca. 45,000 km^2 Ladakh region. We also understand that most park staff lack the necessary clothing, equipment and housing necessary for effective work in the region. The situation in Lahul & Spiti is not very different. It is an urgent requirement for the Centre and State Governments to provision the necessary resources to these areas for effective conservation in the region.

GRAZING COMPETITION BETWEEN WILD AND DOMESTIC HERBIVORES

The entire region has heavy dependence on livestock. While some of the people are agro-pastoralists, many are entirely pastoral. Estimates range from 10 to a few hundred livestock heads per household in the region. Livestock population in the Indian Trans-Himalaya has been growing continuously in the past decades, as evidenced by data available from Ladakh, were the livestock population has almost doubled between 1972 and 1992. Even though conclusive information on habitat degradation, and direct competition between domestic and wild herbivores from the region has just started coming, it is evident from some preliminary studies that the present livestock grazing levels in areas such as eastern Ladakh and Spiti may already be unsustainable. The potential impacts of excessive grazing by livestock include depletion of the scarce forage for wildlife, habitat degradation, disease transfer, and reduction in the breeding performance of both wildlife and domestic stock.

Conclusive studies to ascertain impacts of livestock grazing need to be taken up at many sites. There is also an urgent need to see how the pastoral and agro-pastoral communities of the region can be drawn into a trade-off that reduces their dependence on large livestock holdings, while at the same time helps in improving their standard of living. An example for such an effort was made by the Nature Conservation Foundation, Mysore, and details are given below.

KIBBER GRAZING RESERVE

The Nature Conservation Foundation, a science and conservation organization based in Mysore, signed a written agreement with the village council of Kibber in Spiti in the year 2000, where both these institutions resolved to protect a 5 km2 area exclusively for wildlife by agreeing not to graze their livestock in that region.

The rangeland area has been traditionally used for livestock grazing and collection of fuel, and medicinal plants. Two years of protection is already showing signs of wildlife recovery, as indicated by the increased use of the area by bharal. The compensation costs for lost grazing are being met with by the Van Tienhoven Foundation in the Netherlands, and the project is being implemented voluntarily by scientists associated with the Nature Conservation Foundation and the Wageningen University. The International Snow Leopard Trust has recently joined hands with the initiative, and these institutions are working together towards off-setting the costs that the local people are bearing for living with wildlife (through programmes in conservation education, supporting self managed insurance schemes, value addition to local handicrafts), and towards enabling the local people to benefit from the wildlife they share their resources with (wildlife tourism).

Conflicts Relating to Damage of Crops and Pastures by Wild Herbivores

The Jammu and Kashmir Department of Wildlife Protection staff in Leh has been receiving compensation claims for damage to crops by species such as bharal and urial.

The extent of such damage is not yet clear, but there is an increasing trend in such claims. There is a need to have a better record of conflict cases. 'Hotspots' of such conflict zones should be clearly identified and if any are found, participatory exercises should be taken up to minimize the losses.

In a surprising development some nomads and state Govt. officials claim that the kiang are now damaging the winter pastures of the valuable pashmina or Cashmire goats in Changthang.

It has been argued that such claims are largely baseless and are probably a result of reduced tolerence levels among the people in recent times.

Livestock Depredation by Wild Carnivores

Livestock depredation seems to be a serious conservation issue in the Trans-Himalayan region. As indicated earlier, livestock rearing at present forms an important part of the local economy and any loss to livestock results in a direct monetary loss to the local herders. Park staff in Ladakh report that often up to 60% of their annual outlay goes in meeting the livestock depredation compensation claims filed by people. Damage to livestock takes place in the pastures as well as in the night time corrals. In India's Trans-Himalayan zone only four studies have as yet quantified the extent of livestock damage due to depredation by wild carnivores. These studies are by Mishra (1997) from Spiti, Bhatnagar *et al.* (1999) from the Hemis NP, Ladakh, and by Jayapal (2001) and Sathyakumar (2001) from Zanskar, Ladakh. The damage to livestock in many of these areas is quite high and in some villages up to 14 animals per household have been lost in an year. The monetary loss to households in the Hemis NP averaged ca. Rs. 12,000/-during 1996-97. This study also showed that over 40% of the losses were taking place in the corrals, an aspect that can be dealt with more easily than the damage in pastures. Small and effective means of alleviating these conflicts have been developed by ISLT and SLC, along with local NGOs and the Wildlife Department in Hemis NP, which has potential for replication elsewhere.

Integration of Efforts by Different Government Departments and Non-Government Organizations

Owing to the remoteness of the Trans-Himalayan region, the state governments have resorted to a system of governance that is called 'Single Line Administration'. Under this system, the district head, the District Commissioner (DC) or the Additional District Commissioner (ADC) becomes the head of all Government Departments working in the region. In addition, the Ladakh region, that constitutes bulk of the Trans-Himalayan region in India, has a Ladakh Autonomous Hill Development Council (LAHDC), a form of local Government.

After consultation with the Wildlife Department in Ladakh, it was apparent that inter-agency cooperation and coordination in the region is lacking, leading to inefficient functioning by the Wildlife Department. Examples of major developmental activities being undertaken inside wildlife PAs were cited as cases when the Wildlife Department had to stop such activities when they were well underway. This earned the ire of the local people as well as the respective Government Departments undertaking the work. For these reasons, we feel that the 'Single Line Administration' would facilitate coordination between Departments more effectively, specially if there are relevant policies and practices in place that make it imperative on the respective DCs or ADCs to keep wildlife conservation interest in mind before approval of any development schemes.

The Ladakh region has an added advantage of the existence of numerous non-government organizations (NGOs), many of which have a good reputation of grassroots work in the fields of alternative sources of energy, organic agriculture, and education. This is a resource that should be effectively tapped for conservation related work.

Political Issues

The entire Trans-Himalayan region has international borders. Ladakh has a large and hostile border along the west and northwest with Pakistan held Kashmir called the Line of Control (LoC). On the north and east is the international border and the Line of Actual Control (LAC) with China. Himachal Pradesh and Sikkim also share borders with China. Numerous stretches along these borders are disputed territory. Species such as the Tibetan antelope, argali, kiang and Tibetan gazelle occur at numerous places along and across the border with China. Good snow leopard, ibex and urial habitat occurs along the LoC. Often due to the sensitive nature of the region the Wildlife Departments have little control over the region. There is heavy presence of defence forces on both sides and wildlife of the region might also be a casualty to the frequent skirmishes. The problem is compounded when these may be the only places within the country where a species occurs. Trans-border conservation of these species should thus be given high priority.

Jammu and Kashmir has a separate Wildlife Protection Act, which had placed endangered species such as the Tibetan antelope and brown bear under Schedule II, for which hunting licenses could be given. The amended Act is however at par with the National legislation. The entire Trans Himalayan region spreads across three states and inter-state collaboration in conservation efforts though difficult, is nevertheless necessary. What we want to stress here is that the political dimension is important to consider while planning any large-scale conservation effort in the Trans-Himalayan region, a region replete with sensitive borders spreading over three states.

Poaching of Snow Leopard and Prey Species

Sport hunting was quite widespread during the British period in the Trans-Himalaya. Even after independence the trend continued till the early 1980's when the defense forces, Government officials and others were known to hunt in various parts of the Trans-Himalaya, especially in Ladakh. This had decimated the populations of numerous species in the region. Recently, however, there is evidence of a decline in hunting in many parts of the Trans-Himalaya, with the revival of some wildlife populations in the region such as the Ladakh urial.

Buddhism is the dominant religion in most of the Trans-Himalayan region in India and hunting is generally not practiced, unless it is in retaliation for some damage to their property. However, in western Ladakh and in Lahul, hunting might still be an issue. Again, little information exists on the extent of poaching going on in the region. Some illegal trade in wildlife products, including snow leopard parts and shahtoosh, may be occurring in the region. We understand from the Wildlife Department in Leh that no hunting license has been issued since the mid-1980's. Until trends in poaching and wildlife trade in the region are better documented and understood suitable measures cannot be devised and undertaken to minimize the problem.

Wildlife Diseases

Wildlife disease can be damaging and may even lead to the extinction of small populations.

This is particularly true for small, isolated populations as the Tibetan gazelle in Ladakh, and snow leopard in parts of the Trans-Himalaya. Information on wildlife disease from the region is, however, completely absent and there is an urgent need to generate such information. The reports of infectious diseases such as PPR and FMD in livestock of the region increase the threat of an epidemic in the wild herbivore populations. It is suspected that the frequent imports of livestock from the plains by the armed forces for meat could be a source of exotic diseases in the region. Effective quarantine and screening of all imported animals and vaccination programmes for livestock are needed for the entire region, but on priority for areas in and around the PAs or areas with endangered species.

Some measures that could be undertaken in the region to aid conservation have already been discussed above in the section dealing with conservation issues in the Trans-Himalaya. What we give below is an indicative way of planning wildlife conservation in the region as a whole. We understand that these ideas will have to be fine tuned through wider stakeholder consultations. What we have tried to argue so far is that:

- Even though there are some large PAs in the Trans-Himalaya, bulk of these are occupied by permanent ice and sheer rock faces-the effective areas important for wildlife within them are usually small. Some of them may not even qualify as a PA.
- For existing PAs an effective vision for management is mostly absent.
- Most of the large mammals in the Trans-Himalaya, need large areas given the sparse resources, and seasonal movements. Further, wildlife in this region is mostly outside the existing PAs and simply adding more areas under the PA network may not be a viable solution.
- Areas in PAs usually form an important resource for native people, for whom few alternative livelihood options are available. Traditional concept of large inviolate PAs is not practical in the region.

Based on the conservation issues presented above, and the almost continuous wildlife distribution in the region, we feel that conservation in the Trans-Himalaya has to be planned with a regional perspective, in which the native people are taken as an integral part of the conservation efforts. Having large inviolate national parks and sanctuaries does not seem viable in the region. The shift of focus for conservation in private lands is a need recognized by conservationists worldwide. Using this line of thinking, we feel that continuing with the existing scheme of PAs may not work in the Trans-Himalayas and we now need an alternate paradigm for wildlife conservation in the region. One of the ways of moving ahead is to carefully work on the zonation of existing PAs and of the larger Trans-Himalayan landscape in general.

AN ALTERNATE ZONATION CONCEPT

ZONATION WITHIN EXISTING PAS

PAs in India have zones of varying landuse, such as a core zone, which is inviolate, and a buffer zone that may have multiple-use. The latter zone further may have areas earmarked for forestry operations, tourism and other consumptive uses. Our information suggests that none of the existing PAs in the Trans-Himalaya have cores and buffers delineated. The NPs are essentially 'core zones' in their entirety. With the enhancement in the legal status of the WLS following the 1991 amendment of the Wildlife (Protection) Act, 1972, all the PAs constituting ca. 15,000 km^2 have technically become 'core zones'. Continuing with this practice, as we have already seen, is not pragmatic. We thus suggest that for the existing five PAs in the Trans-Himalaya we change the management zonation approach by carefully delineating core zones in a mosaic, with a buffer area surrounding it. The difference from the existing scheme, primarily is that we do not take impractically large areas of a few thousand km^2 as inviolate core zones and we have multiple core zones within any PA based on its value.

The steps to be followed may be as follows:

- Carefully survey all PAs to determine areas that have high wildlife value, either in terms of presence of an endangered species such as snow leopard, Tibetan gazelle or argali, or in the presence of a large diversity of large mammals.
- Designate such area as a core zone using a participatory approach.
- Have at least one such core area for every 100 km^2 of the PA, although consideration of species movements may need to be included in their planning.
- The peoples rights may need to be settled using innovative schemes such as those outlined.
- The buffer zones would be all the remaining area in the PA where traditional use may continue. Attempts would however, be made to minimize the negative impacts of human use.
- Focused studies that help in determining suitable levels of use need to be encouraged.

In areas of the Trans-Himalaya Outside Existing PAs

The issue of conservation of wildlife outside PAs is more complex. As alluded to earlier, the need for this arises because:

- Endangered species like the snow leopard, Ladakh urial, Tibetan gazelle, argali and chiru occur mostly outside existing PAs
- Including all such areas under the PA network would mean that over 50% of the Trans Himalaya will need to come under the PA network
- Resources for managing such a large network, both material and human are not available
- It is likely to further increase people-park conflicts manifold

There is thus a need to recognize these areas for their importance and device strategies for their conservation. This is with the knowledge that local culture/values and the ow human population density have an important contribution in wildlife persisting in these areas. Zonation seems one clear means to develop this strategy. We realize that aspects of what we present below may be theoretical but will surely be useful in developing a framework. In the alternate approach, we suggest further four landuse zones, which have subtle differences in their management objectives. The first zone is the 'Conservation Zone' which is a small, carefully researched and selected area preferably measuring more than 10 km^2, and where ever possible, up to 100 km^2, where the local people agree to give up their rights in exchange of some development schemes. These are then surrounded by the 'Alternate Livelihood Zone', which would be the most widespread zone. Here the various Government departments' and NGOs work together to limit livestock numbers and dependence on natural resources so as to allow sustainable utilization of resources.

This zone can also have agro based and other non-polluting industries to create employment. The third zone can be 'Low Value Zone', which includes unusable areas under permanent ice and large rock faces. These areas will constitute a major portion of the area. The zone can also include narrow stretches of areas that have lost all value for maintenance of wildlife such as townships and excessively degraded areas.

These zones would go a long way in addressing the issue of conservation and development using a regional perspective.

AREAS WITH INFORMATION GAPS AND INDICATIVE ACTIONS

ADDITIONAL LIVELIHOOD OPTIONS

At present, we have the options of cash crops such as green peas and potato that can be marketed as fresh vegetables in markets in the plains and locally. To minimize the loss through decay during transportation to markets, part of the produce could be processed locally into processed food products or health food. Another industry suggested is electronic industry that is usually less polluting to the environment. But issues regarding power and access need to be addressed first in order to make such ideas viable. The region needs better schools and colleges and has potential for establishment of national level educational institutions. For all these activities, enhancement of the present infrastructure is extremely important. Ladakh already attracts large number of international and domestic tourists. The benefits from tourism are however largely limited to a very small population within Ladakh. There is tremendous potential for development of ecotourism schemes in the region that would enable tourism to take place in a sustainable manner and with substantial benefits reaching local residents. The potential for the development of nature tourism and handicraft based industry should thus be explored as a means of alternative livelihoods. This however, should not be developed as the only means of sustenance of families.

Some highly innovative nature tourism schemes are being developed for Ladakh by organizations such as the Snow Leopard Conservancy, The Mountain Institute along with local organizations such as Ladakh Ecological Development Group (Jackson, Rodney and Jain Nandita, pers. comm.). More effort needs to be devoted to establish optimal stocking densities for livestock in different parts of the range. Grazing competition between livestock and wild herbivores seems to be a significant conservation issue. However, quantitative information on impacts of this is grossly lacking from the region. The studies should also try to determine stocking densities that enable wildlife to exist at levels that allow them to breed and sustain a healthy population. These studies will be crucial for recommendations for the permitted grazing in the multiple-use areas.

Wildlife and People Need to be Established

Data on actual levels of conflicts, the wildlife species involved and conflict 'hotspots' is often lacking from the region. For designing any conflict resolution scheme, such data is of immense importance. With such information, mitigation measures such as corral improvement, in small, but effective ways should then be taken up to resolve the issues. Innovative livestock insurance schemes are also an important possibility. These can be taken up in conjunction with programmes that help in actual reduction of the damage. One such scheme has been designed in Baltistan where the community managed insurance funds are complemented by money generated through wildlife tourism. Another one is being formulated in Kibber wildlife sanctuary, Himachal Pradesh, India by the Nature Conservation Foundation-ISLT. Resolving conflicts effectively will have a two pronged benefit. One is that the monetary loss to the local herders will be reduced and second is that they will be more sensitive to conservation efforts.

CONSERVATION AWARENESS INITIATIVES

Conservation awareness initiatives that illustrate the peculiarities and fragility of the local environment need to be taken up for the local people, tourists and importantly, the district Government officials and politicians. For the latter these may be in the form of relevant directions from various Central Ministries for keeping wildlife conservation perspective in view when developing conservation schemes.

The Wildlife Department, in collaboration with scientific organizations needs to develop Management Plans for the existing PAs with relevant zonations in place. All other potential areas that could serve as the revised 'Core Zones' need to be surveyed immediately. Issues relating to infrastructure need to be addressed to the Ministry of Environment and Forests. The revival and redrafting of the 'Snow Leopard Scheme' could be an ideal opportunity to bring in the suggested changes in the conservation of the region.

The Wildlife Institute of India, Dehradun, along with its partners, the International Snow Leopard Trust and the US Fish and Wildlife Service have already undertaken a step in the direction of generating information on the gap areas and also conservation efforts that enable better trained staff with sound management plans in place. The programme also intends to try and influence policy for conservation in the region.

THE HIMALAYAN REGION: MOUNTAIN ECOSYSTEMS OF THE WORLD

The Indian Himalayan region occupies a special place in the mountain ecosystems of the world. These geodynamically young mountains are not only important from the standpoint of climate and as a provider of life, giving water to a large part of the Indian subcontinent, but they also harbour a rich variety of flora, fauna, human communities and cultural diversity. Despite the abundance of natural resources, most of its people are marginalized and still live on subsistence level. The unsci-entific exploitation of natural resources is leading to increasing environmental degradation and ag-gravating the impact of natural hazards. There is a need to evolve new paradigm to restore balance between economic interest and ecological imperatives with due regards to socio-cultural principles.

With the increasing realization that the natural resources of mountain areas are vital for both upland and downland people, the Global Agenda for sustainable development has brought mountains to sharp focus. Development needs addressing local aspirations and national compulsions have to be met if economic upliftment is to be achieved. However, development interventions also imply a demand on re-sources as well as modifications of existing natural sys-tems. Development in the mountains, therefore, has to have a different approach, given the fragility and vulnerability of the Himalayan ecosystems due to the uniqueness of mountain specificities. Development interventions ignor-ing the imperatives of mountain specificities will invaria-bly result in resource misuse and subsequent accelerated environmental degradation, which would be disastrous not only for the local populace, but also for downstream inhabitants. Such negative impacts of unplanned develop-ment, insensitive to mountain specificities, are already becoming common, the most frequent being the regular incidences of landslides, river obstructions and flash floods in the mountain and recurrent floods in the plains. In addition to the negative impacts of localized develop-ment activities, the effect of climate-induced changes, an outcome of unsustainable practices and waste generation, on the mountain systems is frightening. Global warming and its effects on glacier recession have far-reaching implications both in time and space. The intense vulnerability of mountain ecosystems and their elements to the human as well as climate-induced changes, therefore, is of great con-cern. Not surprising, therefore, that the complexity of such issues continues to receive considerable attention at the global fora like the WSSD (World Summit on Sustainable Development, Johannesburg, August 2002) and Bishkek Global Mountain Summit (October 2002). These events have arrived at a consensus that mountains would require specific approaches and resources for sustaining livelihood needs and improving the quality of life. This would require an integrated approach, which gives due consideration to closely intertwined aspects of human socio-cultural/socio-economic systems and natural ecosystem components/processes.

UNIQUENESS

Among the global mountain system, the Himalaya is the most complex and diversified, and separates the northern part of the Asian continent from

South Asia. The region being a discrete geographical and ecological entity, figures prominently in major biophysical settings of the planet earth. This vast mountain range (over 2500 km in length, between 80 and 300 km wide and rising from low-lying plains to over 8000 m asl) produced a distinctive climate of its own and influences the climate of much of Asia. The great variation in topographical features causes immense diversity in climate and habitat conditions within the region. Tem-poral and spatial variations caused by diversity in geological orogeny have resulted into a marked difference in climate and physiography and consequently in the distribution pattern of biotic elements. This spatial position and heterogeneous dispersion of biodiversity elements have led to complexity in biogeographical patterns of the region. The eastern Himalaya (including northeast India) that harbours about 8000 species of flowering plants is considered a cradle of flowering plants, whereas the western Himalaya supports over 5000 species of flowering plants. The Indian Himalayan region (IHR) as a whole, supports nearly 50% of the total flowering plants in India of which 30% flora is endemic to the region.

There are over 816 tree species, 675 edibles and nearly 1743 species of medicinal value found in the IHR. In view of growing threat to biological diversity, conservation and rational use of biodiversity in the Himala-yan region could bring enormous economic benefits to the local populations and can indeed contribute to sustainable development.

The region is known as a 'water tower of the earth'. Approximately 10-20% of the area is covered by glaciers, while 30-40% remains under seasonal snow cover varying from 0.48 0.43 to 2.20 1.25 million km^2. Despite the vast water resources (1,200,000 million m^3 annual flow of Himalayan rivers) trends such as diminishing regulatory effects of glaciers, streams and rivers are gradually occurring in the region. This region has a total geographical area of about 530,795 km^2 inhabited by 31,593,100 people, representing 16.16% of the total area and 3.73% of the total population of India. The literacy rate (7 years and above) of IHR (about 67%) is marginally higher than the national average (65.4%) recorded in the 2001 census. Its forests display phenomenal biodiversity that is used to meet diverse needs

of the people. The forest biomass value in some oak forest stands of Central Himalaya, 545-782 t ha^{-1} yr^{-1}, is typical for the region. The Himalaya with its vast green cover acts as 'sink' for carbon dioxide. Estimates of annual carbon sequestration by the forests of western and northeastern Himalaya are computed to 6.49 mt, that values to 843 million US$ (A. K. Tewari, unpublished data). This is one of the important ecosystem services being performed by the Himalayan forests. The beautiful landscapes, numerous rivers and streams cascading down the mountain slopes, diversity of cultures and religions, and colourful festivals of indigenous/ethnic communities present strong attractions for people from all over the globe, be they nature-lovers, tourists, or seekers of peace and truth.

ENVIRONMENTAL SECURITY AND PEOPLES' ASPIRATIONS

The people of the IHR, like elsewhere in other mountain ecosystems, are heavily dependent for their livelihood on their immediate natural resources and production from primary sectors such as agriculture, forestry, livestock, etc. The dependency of the continually growing population on finite resources, lack of viable technologies to mitigate the mountain specificities and enhanced production to meet the demands are depleting the resources along with increasing marginality of farmers, ultimately promoting poverty.

Despite its rich biological and cultural resources, the region is under-developed. Present trends of environ-mental health suggest that existing interventions are un-sustainable. Economic indicators also do not reflect the desired effects on economic upliftment.

In addition, the inherent fragility of the mountains as well as the increased vulnerability of the Himalaya to human-induced environ-mental impacts make people live in the shadow of fears of natural hazards. Large number of studies carried out in the region focusing on development interventions/initia-tives reflect the unscientific exploitation of resources lead-ing to increasing environmental degradations. Reduced dense forest cover, accelerated soil erosion and increased silting of water bodies, drying-up of springs, re-placement and disappearance of species and increased ratio of energy expended in fodder, fuel collection, and agricultural activity that increase drudgery of the womenfolk are some of the tell-tale symptoms of environmental ill-health.

ENVIRONMENT AS A HOLOCOENOTIC RESOURCE SYSTEM

Ecology is the science that elicits the functional inter-relationships among the different components of environ-ment on the one hand, and between the organisms and environment on the other. A major ecological principle states that the environment is holocoenotic in nature, and therefore any change in one component is bound to change the states of all other components.

For example, deforestation leads to increased run-off (hence floods), increased soil erosion (hence siltation of water bodies), disappearance of species (hence gene erosion), and atmospheric loading of CO^2 (hence global warming).

Thus the demand for timber and firewood across the country has had an impact on the forests of the Himalaya and deforestation in the Himalaya affects the flood situation in the Gangetic Plains. This ex-plains how the scale of deforestation effects ranges from local to regional to global.

Thus environment not only com-prises the life-support system for biological organisms, but is also a system of interacting resource subsystems. The term resource implies management. Proper manage-ment will not disrupt a system because a dynamic equilibrium will be maintained among its subsystems and components. Therefore, environmental degradation is the outcome of mismanagement leading to imbalance and over-exploitation of resources.

Interdependence of ecological and socio-economic activities The problems in the Himalaya are complex, having intri-cate linkages between social, economic and ecological concerns.

The solutions, therefore, cannot be addressed in isolation. To cite an example, the agro-and forest ecosystems are so intricately inter-related and inter-dependent that it is futile to talk of forest management in isolation without considering the cropland. In the Central Himalayan region, it is estimated that the cost of subsistence agriculture on the forest ecosystem is high. For example, for each unit of en-ergy obtained in agronomic production, seven units of energy are expended from the forest through the use of firewood, fodder and vegetal manure. A greater ratio of forest to cropland (5.18: 1) is needed for sustenance of agriculture against the present ratio of 1.66: 1. Problems have appeared because of the reduction in this ratio, im-plying that the carrying capacity of forests has already been exceeded.

Similarly, pasture development or revege-tation of wasteland without solving the problems of animal husbandry, fodder and fuel is not possible. The traditional agri-silvi-pastoral mode of subsistence living of the inhabi-tants of the region is no more sustainable, both ecologi-cally and economically.

It is apparent that sectoral practices of management (or development) will not work, and therefore, the only approach which will work is a holistic one consistent with ecologi-cal and social principles.

This approach also implies that the hill and adjoining plains must be taken as the macro-planning unit, with smaller structurally and functionally definable units for micro-level planning. The various eco-systems should be categorized into protective, productive and waste-dissipative systems and should be managed accord-ing to their roles.

Therefore, the basis of any planning for sustainable development in mountain areas has to be centred around man's relationship with nature. The rela-tionship is desired to be governed by a sense of justice and equity. Each culture is the result of the people trying to survive within their environment and indeed of an at-tempt to optimize the use of its resources.

Lifestyle and production systems develop steadily by experimentation and observations over centuries, till they become so cul-turally incorporated that they are like genetic knowledge.

This has been inherent in many tribal societies, but in the modern acquisitive society 'economy' gets priority over 'ecology'. There is need to evolve a new paradigm to re-store balance between economic interests and ecological imperatives. Although the ecological and economic sys-tems have a myriad of inter-connections, the most simple and most obvious is this: ecological system provides raw materials to the economic system and absorbs the waste generated by the economic system.

Therefore, the system will be constrained by the productive and waste-absorption capacities of the ecological system. When one or both these capacities are exceeded, ecological backlashes are bound to occur.

Once the waste-dissipative capacity of the Ganga was exceeded, severe pollution problem emerged which is now costing the government a huge sum of money, with still doubtful level of final out-come. Similar is the case for water bodies in the hills. When timber extraction or biomass extraction exceeded the limit of harvestable productivity of the forest, the lat-ter began to diminish.

The ecological and economic considerations are there-fore to be combined to attain ecologically sustainable deve-lopment. Both ecological and economic values can be served individually in a variety of ways, but combining ecological and economic considerations adds geometri-cally to the complexity of development programmes.

When socio-cultural systems are added onto the ecological-economic relationships, the situation becomes further complicated. However, development driven solely by economic considerations has changed the aspirations, value systems and management priorities. Demographic and legal factors further complicate the application of ecological considerations to development of goals and processes.

ACHIEVING SUSTAINABLE DEVELOPMENT

Simply stated, sustainable development implies the use of ecological system in a manner that satisfies current needs without compromising the needs or options of future gen-erations. Strategies for sustainable development must be based on reliable and comprehensive data on natural, socio-cultural and socio-economic resources, as well as on the environmental set-up.

These strategies should incorporate traditional knowledge and established production systems after they have been carefully evaluated. The aim of sustain-able development should be to maximize human well-being or quality of life without jeopardizing the life-support environment. Although there is no unique defini-tion of quality of life, the following groups of variables together might be considered its indicators:

(i) economic variables: per capita income, employment stability, income distribution;
(ii) ecological variables: ecological degradation, environmental quality, use of renewable and non-renewable resources, human-initiated energy con-sumption;
(iii) social variables: social security, emotional support, intellectual growth and mental satisfaction;
(iv) cultural heterogeneity, and
(v) political variables: scope and use of government services, political participation, political power advantage and policies.

Solutions to the ecological and economic problems in the IHR are to be sought within the permissibility of mountain specificities and adaptability of people, which is governed by socio-cultural principles.

Identifying sustainable landuse practices, promotion of on-farm activities, value addition to all resources and adoption of environment-friendly technologies, restoration of degraded ecosystems, biodiversity conservation, water resource and hydro-power development, promoting community-based management, upgrading in-frastructure, improving quality education and capacity building to ensure benefits, are but a few priority activities to improve livelihoods, income and environment of the IHR.

CONCLUSION

The unique majestic Himalaya has provided immense ecosystem goods and services in the past and, with proper planning and management will be able to provide the same in the future also. However, we must acknowledge the fact that the whole IHR is facing anthropogenic pressure leading to overall degradation of its environment. Once symptoms of environmental deterioration become apparent, most often the only option left is to react to the situation and try to cure the problems by costly corrective measures.

It is much better however, to be able to anticipate the problem and take up preventive measures in the beginning. Proper education at various levels, long-range database and a holistic approach would bring us nearer to sustainable development ensuring better quality of life, improved eco-nomic status, and minimized adverse effect on life-support environment. There is also a need to experiment and devise ways and means to ensure that the development does not destroy its bio-cultural diversity and social fabric. The Himalayan region should not be burdened with back-ward-dragging heritage of the past, nor be constrained by the mistakes that bigger states have committed.

7

Forest Hydrology

INTRODUCTION

Forest hydrology combines aspects of two separate disciplines: hydrology and forestry. Hydrology is the science that studies the waters of Earth. Hydrology seeks to understand where water occurs; how water circulates; how and why water distribution changes over time; the chemical and physical properties of water; and the relation of water to living organisms. Forestry is the science that seeks to understand the nature of forests and the interactions between the parts comprising a forest. Forest management started in the 1700s as a form of large-scale farming to improve yields of timber and fiber from forests. In the United States, watershed protection has been an integral part of forest management since its origins. The Organic Administrative Act of 1897 stated that forest reserves were to protect and enhance water supplies, reduce flooding, secure favorable conditions of water flow, protect the forest from fires, and provide a continuous supply of timber. The 1911 Weeks Act authorized the acquisition of federal lands in the East for the express purpose of protecting the watersheds of navigable waterways. In recent decades, forestry has adopted more of an ecosystem management approach while still including timber production as an important goal. Although a forest is an ecosystem dominated by trees, a healthy forest includes other plants as well as soil, terrestrial and aquatic animals, and water—plus people who use the forest and its resources. Modern forest management therefore requires not only an understanding of forest science, soil science, and hydrology, but also principles of wildlife biology, land-use planning, and recreation planning.

WATER AND FORESTED ECOSYSTEMS

Ecologists consider water to be the defining part in an ecosystem, including the forest ecosystem. Water shapes the physical landscape through erosion Undulating forest surrounds meadows, lakes, and rivers of the Charlevoix Conservation Area, a coastal watershed in Quebec, Canada. This region, which illustrates a high-quality forested watershed, has been designated a biosphere by the United Nations Educational, Scientific and Cultural Organization (UNESCO). and deposition. It also shapes the biological parts of the ecosystem by its presence or absence; its quantity and quality; and its occurrence and distribution. The water cycle plays a key role in ecosystem functions and processes. Forests, in turn, are vital to the water cycle and to water quality. In essence, the forest acts like a giant sponge, filtering and recycling water. Approximately 80 percent of U.S. fresh-water resources are estimated to originate in forests, which cover one-third of the U.S. land area.

Tree leaves intercept water from rain, snow, and fog; the leaves also release water back to the atmosphere by evapotranspiration. Tree roots extract water from the soil while helping hold the soil in place. Forested land reduces the surface impact of falling rain through interception and delay of water reaching the surface. Forestland also decreases the amount and velocity of storm runoff over the land surface. This in turn increases the amount of water that soaks into the ground, a portion of which can ultimately recharge underlying aquifers. Conversely, water from hydraulically connected surficial aquifers may enter streams and wetlands, helping to maintain their water levels during dry periods.

FORESTS AND THE HYDROLOGIC CYCLE

The surface water in a stream, lake, or wetland is most commonly precipitation that has run off the land or flowed through topsoils to subsequently enter the waterbody. If a surficial aquifer is present and hydraulically connected to a surface-water body, the aquifer can sustain surface flow by releasing water to it. In general, a heavy raïnfall causes a temporary and relatively rapid increase in streamflow due to surface runoff. This increased flow is followed by a relatively slow decline back to baseflow, which is the amount of streamflow derived largely or entirely from groundwater. During long dry spells, streams with a baseflow component will keep flowing, whereas streams relying totally on precipitation will cease flowing. Generally speaking, a natural, expansive forest environment can enhance and sustain relationships in the water cycle because there are less human modifications to interfere with its components. A forested watershed helps moderate storm flows by increasing infiltration and reducing overland runoff. Further, a forest helps sustain streamflow by reducing evaporation (*e.g.*, owing to slightly lower temperatures in shaded areas). Forests can help increase recharge to aquifers by allowing more precipitation to infiltrate the soil, as opposed to rapidly running off the land to a downslope area.

Riparian Areas

The riparian zone is broadly defined as the area between a body of water and the upland parts of the landscape that are rarely flooded except under the most extreme conditions. But the term also can refer more specifically to the immediate streamside area. Riparian areas represent less than 10 percent of most forest ecosystems, yet these areas often are the most productive portions. Compared to upland regions, riparian areas have more water available; the vegetation is more robust; the soils are deeper; the timber often is of higher quality; and the waterbodies have more shade. The riparian zone also may include wetlands bordering streams and lakes. This combination of factors makes riparian areas among the most heavily used portions of a forest. Riparian and wetland areas provide abundant and reliable forage for wildlife, as well as transportation corridors. They also may receive heavy human use for recreation.

Riparian zones also are attractive destinations for logging and for livestock grazing; as a result, riparian areas in forests are sometimes heavily damaged, especially in the forests of the arid American Southwest. Fortunately, riparian areas respond well to good management practices.

Aquatic Biodiversity

Forest lands and waters are vitally important in maintaining biodiversity and providing habitat for fish and wildlife, including threatened or endangered aquatic species. In the United States, over one-third of national forest lands are critical for maintaining aquatic biodiversity and protection of listed species. For aquatic species, watersheds provide the basic unit of any conservation strategy. Many watersheds also contain isolated habitats with unique characteristics producing a high potential for rare species. Some species occur only near a single spring or in a single stream within a given watershed. Lands set aside to protect these unique habitats also benefit the entire watershed and its ecosystem.

FOREST MANAGEMENT AND WATERSHED QUALITY

Wind, fire, insects, and disease are all part of properly functioning, healthy ecosystems in watersheds. For example, natural fires, although temporarily devastating, periodically restore the balance between vegetation types, and release nutrients from the vegetation and soil. In contrast, widespread clear-cut logging and excessive or improper road-building can degrade watersheds, Rill and gully erosion caused by flowing water is evident on this steep slope following a forest fire. (Note the person crouching.) The soil loss can be substantial, and denuded slopes can be difficult to revegetate. as can land uses such as ski runs and housing projects. Many human activities can increase overland runoff, resulting in more erosion of the land surface and concurrently reducing the amount of water that soaks in the ground to potentially reach nearby streams or recharge underlying aquifers.

Moreover, fire prevention and suppression have created "imbalanced" forests with excessive amounts of undergrowth and dead vegetative matter that serve as fuels when fire does occur. Hence, these forests are at increased risk of high-intensity, destructive fires. Watershed management and restoration may include controlled thinning, prescribed burning, and other management practices to restore the proper balance of timber, undergrowth, and grassy meadows in the watershed. Restoration also may include planting of appropriate native plants.

Effects of Roads

Improperly engineered roads in forests can increase erosion and significantly increase the risk of landslides. Both of the adverse effects are more severe when roads are numerous, and when they either cross or run parallel to streams. For example, heavy precipitation in Oregon and Washington during the mid-1990s resulted in many landslides, with a correlation between the slides and the frequency and density of roads. With respect to erosion and sedimentation, water runoff flowing along and across roads picks up sediment, which can then be deposited in nearby lakes and streams. This siltation can degrade or destroy habitat for aquatic organisms that require clear water and silt-free benthic (bottom) substrates. Proper road engineering and following good practices (such as the U.S. Forest Service Guidelines for Best Management Practices) can reduce or eliminate the risk of erosion, landslides, and stream degradation due to excess siltation. Unfortunately, many roads in U.S. national forests were built before the practices were in place.

Effects of Fire

Destructive fires that remove large amounts of organic matter in a forest cause loss of nutrients from the soil as the detrital cover (*i.e.*, dead and decaying materials on the forest floor) and upper soil layers are burned and eroded. Moreover, fires can adversely affect the quality of streams and lakes in the burned region as well as tributary watersheds downstream. Surface runoff that otherwise would have been slowed or absorbed by living and dead vegetative matter on the forest floor now runs unimpeded down bare (or nearly bare) slopes. The increased velocity carries more soil particles and loose vegetative matter downslope, along with any adsorbed nutrients.

FORESTS AND 'OCCULT' PRECIPITATION

Therefore the present discussion does not focus on the extent to which a forest cover generates its own rainfall (and over what scale) but rather on the importance of fog and other forms of cloud moisture ('occult' precipitation) as an additional input of water. Wherever fog impacts a forested area, particularly where the fog occurs in the form of an orographic cloud belt or advective coastal fog, additional moisture is intercepted by plant surfaces (or indeed any other obstacle) and precipitation may occur in the form of 'occult' precipitation or 'fog drip', even if no rainfall is recorded on adjacent open ground.

An extreme example has been described by Aravena *et al.* (1989) whose isotope studies have demonstrated that the frequent occurrence of advective sea fogs along the arid coast of northern Chile has given rise to a patchy forest that, in the almost complete absence of ordinary rainfall, thrives primarily on fog. Several studies of fog drip beneath trees along the US Pacific Coast have shown that the amount is directly related to the area and density of the tree profile and to the degree of exposure of the trees to windblown fog. During times of fog, fog drip at the edge of a forest may be several times the rainfall in the open. However, amounts often decrease sharply towards the forest interior.

Therefore, reports of increased net precipitation beneath vegetation need to be considered carefully in terms of gauge positions with respect to the edge of the forest. At favourably exposed locations the extra inputs stripped from the fog by trees (conifers especially) may be considerable. Examples include 425 mm in 46 rainless summer days below 18-m tall Douglas fir in northern California, 760 mm annually measured in Hawaii below *Araucaria* by Ekern (1964) and 880 mm annually for tall old-growth Douglas fir in Oregon by Harr (1982). Bruijnzeel and Proctor (1995) and Bruijnzeel (2000) recently reviewed the hydrological characteristics of 'cloud forests' found in orographic cloud belts on wet tropical mountains and reported extremely variable net precipitation fractions for these forests (55-130% of incident rainfall).

As such, although tropical montane cloud forests are supposedly exposed to frequent fog, the associated amounts intercepted by the trees are not always sufficient to raise net precipitation totals significantly above the 70-80% commonly recorded for montane rain forests on cloud-free locations. Such variations may be interpreted in terms of exposure to and persistence of fog, as well as to contrasts in canopy epiphyte biomass.

Quantification of the amount of fog intercepted by a forest is difficult, particularly if the fog occurs together with rain. The usual approach is to compare amounts of net precipitation (often throughfall only) beneath the forest with the rainfall measured in the open or to subtract the latter from the catch obtained with some kind of fog gauge placed next to an ordinary rain gauge. A number of different fog gauge designs have been proposed, including wire harps, cylindrical screens, 1m^2 polypropylene screens and louvered metal gauges.

The through-fall method essentially gives an estimate of 'net' fog drip since it includes an unmeasured amount of water lost to evaporation from the wet vegetation. In addition, the result is site specific.

The use of a fog gauge (regardless of the type adopted) also requires a site-specific conversion factor to relate the catch of the gauge to that of the actual forest canopy. Some progress has been made with physical fog deposition models but the data requirements of these models are such that their use at remote forest locations must remain limited to a few well-researched sites.

Therefore there may be some merit in alternative approaches to the quantification of fog contributions, such as the sodium/chloride budget method or the contrasting stable-isotope signatures of rain and fog water.

For example, by comparing the stable isotope compositions of rain, fog, soil moisture and groundwater with those of xylem water in the dominant species of a Californian redwood forest, Dawson (1998) was able to demonstrate the importance of fog to the different plant groups. During a year with average climatic conditions, 25-50% of the water uptake by the tall redwood trees was shown to consist of fog whereas these figures roughly doubled during a drought year. Under-storey herbs were even more dependent on fog water: 19-50% in an average year and 28-100% in a drought year.

Transpiration (£$_t$) ceases whenever the canopy is fully wetted. Because of this and the reduced radiation and high relative humidity normally associated with such conditions, E_t of trees exposed to frequent fog is low. Together with the additional inputs via occult precipitation, this makes for an especially favourable water balance, *i.e.* amounts of stream-flow or groundwater recharge will be comparatively high. For these reasons Zadroga (1981) expressed the fear that clearing tropical montane cloud forest could well result in substantially reduced water yield, especially during the dry season. Some support for this contention could come from the contrasting dry-season flows emanating from two pairs of cloud-forested and cleared catchments in Guatemala and Honduras reported by Brown *et al.* (1996).

However, both catchment pairs were too different in size and elevational range (and therefore exposure to fog and rainfall) to be sure that the approximately 50% reduction in flow was entirely due to the replacement of the forest by vegetable cropping.

A more convincing case was provided by Ingwersen (1985) who observed a (small) decline in summer flows after a 25% patch clearcut operation in the same catchment in the Pacific North-west region of the USA for which Harr (1982) had inferred an annual contribution by fog of about 880mm. The effect disappeared after 5-6 years. Because forest cutting in the Pacific North-west is normally associated with strong increases in water yield, this anomalous result was attributed to an initial loss of fog stripping upon timber harvesting, followed by a gradual recovery during regrowth.

Interestingly, the effect was less pronounced in an adjacent (but more sheltered) catchment and it could not be excluded that some of the condensation not realized in the more exposed catchment was 'passed on' to the other catchment. The observation of Fallas (1996) of enhanced throughfall in patches of forest surrounded by pasture in the montane cloud forest belt of northern Costa Rica is probably pertinent in this respect. Further (process-based) work is needed to decide on the importance of occult precipitation for water yield from fog-affected areas.

FOREST HYDROLOGY AND MANAGEMENT

Forest hydrology has built a strong foundation of general principles concerning the direct effects of forest management on hydrologic processes from plot studies, process studies, and watershed experiments. The challenge now is to apply these principles to predict how hydrologic processes will respond to many forms of change in forest landscapes. Forest hydrologists have long recognized the need to understand indirect and interacting effects of forest management at much larger spatial scales and longer temporal scales than is possible in plot studies, process studies, and watershed experiments. Indirect effects are responses to forest management that are displaced in time or space, such as fire suppression leading to insect outbreaks that affect forest hydrology. Interacting effects occur when two or more management practices coincide, such as when post-salvage logging and road building have a different collective effect on forest hydrology that differs from their individual effects.

This chapter examines the research challenges faced by forest hydrology as it moves from principles to prediction at larger spatial scales, at longer temporal scales, and in a changing social context. The chapter concludes by outlining the potential for improved cumulative watershed effects analysis that could provide the predictions needed by forest and water managers in the twenty-first century.

SPATIAL RESEARCH NEEDS

A key unresolved issue in forest hydrology is how to apply the findings of hydrological studies in one area to a different area or how to scale up the findings to large watersheds and landscapes. Although forest hydrologists have confidence in the general principles of hydrologic responses to forest management and disturbance, they cannot predict precisely how forest management will affect hydrologic processes in specific places other than those that have been intensively studied. Most forest hydrology studies are conducted in small watersheds that are instrumented to measure streamflow and other hydrologic properties. However, the sum total of the area studied by forest hydrology is only a tiny fraction of the watersheds in the United States, and hydrologists recognize the need to extend hydrologic knowledge from "gauged basins" (watersheds that have measured records) to ungauged basins. Predictions are most needed in ungauged places to better understand hydrologic effects where conflicts sometime arise: in water supply systems for agriculture and cities; rivers where endangered and threatened aquatic species occur; large water bodies such as the Chesapeake Bay; and many, many others.

Forest hydrology is adopting a landscape perspective to examine spatial patterns of forests and associated hydrologic processes and to link principles from plot- and small watershed scales to predictions at larger spatial scales (hundreds of square kilometers). Within a watershed, forests can be located in the headwaters, along riparian corridors, in woodlots in agriculture lands, and in urban or suburban areas. Based on its intra-basin position, a forest fulfills various water-related functions with respect to water quantity and quality. For example, forests in headwaters influence water yield and the quality of water delivered to downstream areas. Riparian forests located along streams throughout a watershed provide key functions for protecting streams from inputs of sediment, nutrients, and herbicides; provide wildlife habitat for terrestrial and aquatic organisms; and support a diversity of other functions. Riparian forests have been greatly altered by economic development, and they are the focus of many forest management guidelines designed to preserve water quantity and water quality.

Hydrologists use models to predict water quantity and quality in watersheds where there are no measured records. Since 2004, a working group for Prediction in Ungauged Basins (PUB) of the International Association of Hydrological Sciences has developed methodologies for assessing uncertainty in hydrologic predictions arising from uncertainties in landscape properties and climate inputs, choice of model structure, and methods of information transfer from gauged to ungauged watersheds. Most hydrological models are developed and tested for gauged basins and subsequently are validated and applied to ungauged areas.

However, models that have been fitted to data in small, gauged watersheds often provide inaccurate or imprecise predictions when they are (1) extrapolated to other small forested headwater basins, (2) extrapolated to future time periods, or (3) applied to large watersheds. This problem of prediction in ungauged basins has preoccupied hydrology researchers for several decades, and is compounded by a lack of information about how direct hydrologic effects interact under the multiple sets of specific conditions that occur in changing forest landscapes. By examining forest hydrologic processes under a wide range of conditions, landscape-scale studies could provide data and understanding to help extend basic forest hydrology principles to make predictions needed by water managers.

Road networks are a pervasive feature of forest landscapes. The location and density of roads in a watershed can influence the hydrologic effects of roads. In many forested areas, roads are concentrated in the valley bottoms immediately adjacent to streams, meadows, and wetlands. These roads have a particularly high potential for delivering runoff and sediment to streams, lakes, and aquatic ecosystems. The legacy of midslope roads from past logging practices is also of concern, because these have a high potential for subsurface flow interception, connectivity to the drainage network, and initiating shallow landslides. Considerable research effort has been devoted to modifying road design and management to mitigate erosion, and to developing techniques to decommission roads. The hydrologic effects of road networks at large scales, and the effects of road decommissioning are not widely studied.

TEMPORAL RESEARCH NEEDS

Water management systems have been designed and operated under the assumption that hydrologic variables such as annual water yield, while varying over time, can be predicted reliably based on instrument records. However, increased understanding of long-term variability and trends in climate has undermined this assumption. If precipitation and streamflow vary over the long term, future annual water yields may fall short of the levels that water supply systems—and attendant agricultural and urban development—were designed to provide. Given this context, it is critical for forest hydrologists to extend beyond general principles to make predictions of how forest management and disturbance affect hydrologic response on time scales that exceed those of most forest hydrology science.

Some forest and stream management plans now include the historical range of variability, which presupposes that (1) past conditions and processes provide context and guidance for managing ecological systems today; and (2) disturbance-driven spatial and temporal variability is a vital attribute of nearly all ecological systems. The historical range of variability helps characterize the variation in quantity, quality, and timing of streamflow from forests. It can also be used to establish baselines for assessing change in water from forests over time.

Forests and their associated hydrologic processes change on time scales ranging from decades to hundreds, or even thousands, of years. The temporal context for understanding forest hydrologic processes involves expanding the temporal scale into the past to consider the effects of past forest practices and into the future to project and anticipate changes in land use and climate. Many different kinds of legacies of past human activities affect forests in the United States. In some areas, native forests have been converted to agricultural and urban uses, and forests have regrown on abandoned agricultural lands in others. Roads and expansion of urban areas have fragmented forests into smaller, less-contiguous patches and created new drainage patterns. Fire suppression has changed the structure and community composition of many forests, especially in those with otherwise active fire regimes. Exotic species introductions, grazing by domestic animals, predator eradication, and timber harvesting methods have changed forest cover and species composition as well. Future urban and suburban development and climate change are expected to continue to alter forest cover and species composition. Human activities will modify forests in the future, and future legacies will reflect current, regional forest histories.

HYDROLOGICAL EFFECTS OF FOREST MANIPULATION

Within the limitations imposed by flow measuring techniques, the practical overall influence exerted by forests on hydrological processes is most clearly illustrated through comparison of streamflow amounts emanating from areas with contrasting proportions or types of forest. Because of the complications related to local climatic contrasts (rainfall, exposure to radiation and air streams) and ungauged subterranean transfers of water from one catchment to another, such direct comparison of results obtained for catchments with contrasting covers can be problematic.

The classical answer to these problems has been the paired catchment approach in which streamflows from two (often adjacent) catchments of comparable geology, exposition and vegetation are expressed in terms of each other (using regression analysis) during a 'calibration phase'. Once a good calibration relationship has been established, one of the catchments is subjected to manipulation of its cover (*e.g.* strip cutting, clearfelling or afforestation) while the other remains undisturbed as a 'control'. Throughout this 'treatment' or 'experimental' phase, streamflow continues to be monitored and any effects of the treatment are evaluated by comparing actually measured flows from the manipulated catchment with those predicted by inserting the flows from the control catchment into the calibration relationship.

Although a more rigorous comparison between catchments is obtained in this way, an underlying assumption is that differences in leakage between the two catchments remain unchanged with time, regardless of catchment cover status. Also, to avoid overstretching of the calibration statistics during the treatment phase to accommodate extremes in streamflow resulting from excessive rainfall or drought, it is important that the calibration period includes sufficient variation in rainfall.

This, of course, renders the paired catchment method a time-consuming (>5-10 years) and expensive affair. In addition, the method is essentially a black-box technique requiring additional process research to reveal the relative importance of different causative factors to explain the obtained results. All this, plus the limited resolution afforded by the paired catchment approach (usually more than 20% cover change is required for effects on streamflow to be detectable), has led to a general decline in the number of such studies in the last two decades and a gradually greater emphasis on physically based modelling.

In the following sections the hydrological effects of:

(i) forest manipulation (thinning, selective logging, undergrowth or riparian vegetation removal, clearcutting followed by regrowth or conversion to other land uses) and

(ii) (re)forestation are reviewed from the experimental manipulation and modelling perspectives.

Effects of forest thinning on rainfall interception

It has long been recognized that amounts of throughfall tend to be inversely related to the stocking of a forest, be it in the context of thinning or logging operations or forest age. However, the steady increase in rainfall interception with age observed for coniferous forests is not paralleled in deciduous forest. Helvey and Patric (1965) reported that E_i in 11-year-old oak coppice in the south-eastern USA did not differ from that observed in 30-year-old and 50-year-old mixed poplar-hickory forests in the same area.

Apparently, the oak coppice had already acquired leaf biomass and roughness characteristics similar to those of the much older forests. As indicated earlier, interception by deciduous forests during the dormant season is lower than during the growing season. Usually, an increase in throughfall of 5-10% occurs during winter. This increase is by no means proportional to the reduction in LAI, which can be up to sixfold. Similarly, the observed decrease in rainfall interception following forest thinning is often not commensurate with the degree of canopy opening. For example, a 50% reduction in basal area of a Douglas fir forest in France resulted in only a 13% drop in E_i. An even smaller effect was noticed in a regenerating eucalypt forest in south-eastern Australia, where E_i was reduced by as little as 4% following a uniform thinning operation that removed 50% of the overstorey biomass.

Likewise, a 13-fold reduction in basal area (corresponding to a change in planting interval from 2 x 2 m to 8 x 8 m) in a dense Sitka spruce plantation in Scotland was accompanied by only a 3.7-fold reduction in E_i who explained this phenomenon not so much in terms of decreases in canopy cover *per se* but rather as a function of a gradually decreasing turbulent exchange between the trees and the surrounding air after opening up the canopy.

Evidence for this contention came from a strong negative correlation between rainfall interception and aerodynamic resistance (r_a), with r_a increasing linearly as the spacing between the trees increased up to 8 x 8m.

Research results from the humid tropics largely confirm the above findings. Veracion and Lopez (1976) and Florido and Saplaco (1981) found negligible increases in throughfall after thinning 30-50% of the biomass of 10-15-year-old and 30-year-old natural pine forests in the Philippines.

Significant increases were obtained only after 70% of the trees had been removed. Ghosh *et al.* (1980) applied a 20% thinning treatment to a broadleaved *(Shorea robusta)* forest in northern India and found that throughfall increased from 72 to 81%. However, because of a concurrent reduction in stemflow of about 4% (and not necesssarily related to the thinning), the overall change in intercepted rainfall was a modest 5%.

Recently, Asdak *et al.* (1998) compared rainfall interception from undisturbed and heavily logged lowland rain forest plots in central Kalimantan, Indonesia. Interception from the logged forest was 6% of gross rainfall compared with 11% in the unlogged forest. However, because the comparison did not pertain to the same forest before and after logging, it is not possible to assess to what extent the observed reduction in E_i reflects the change in canopy cover or pre-existing structural differences between the two stands.

Effects of Thinning and Selective Logging on Transpiration and Water Yield

Whilst the effect of thinning on E_i is rather limited, effects on soil water (and ultimately streamflow) are even smaller. Opening up of a stand not only enhances the penetration of radiation to the understorey vegetation and the forest floor, but also the remaining vegetation will start competing for the extra moisture supplied by the initially increased throughfall. The magnitude and the duration of such effects will differ among locations, depending on the vigour of overstorey and understorey vegetation, climatic conditions (including site exposure) and the configuration of the cutting.

No changes were detected in the streamflow from a deciduous hardwood forest catchment of south-easterly exposure at Coweeta (southeastern USA) after selective logging removed 27% of the basal area; only a 4.3% increase occurred after a 53% selective cut. Also, removing the entire understorey (representing 22% of forest basal area) from a 28-ha catchment of northwesterly exposure in the same area produced an equally modest change. Similarly, in a Douglas fir forest on the west coast of Canada where substantial soil water deficits develop over the summer, Black *et al.* (1980) observed 'remarkably similar' transpiration rates for unthinned stands with little or no salal *(Gaultheria shallon)* undergrowth and thinned stands with a well-developed understorey. In a related study, Kelliher *et al.* (1986) reported that following removal of the salal transpiration by the Douglas fir trees increased by 30-50%, with the greatest increases noted for the plots where the leaf biomass of the salal had been highest.

The overall effect of the removal of the undergrowth on soil water content was therefore a mere 1-3% increase. In the northern UK, Whitehead *et al.* (1984) compared the transpiration patterns of two 40-year-old Scots pine plantations of similar average height but with a more than fivefold difference in stocking. Transpiration in the widely spaced plantation averaged 67% of that of the denser stand, reflecting differences in canopy resistance and LAI at stand level. However, relative transpiration rates *per tree* were 3.3 times higher in the thinned plot and were intermediate in magnitude between the relative increases in average basal sapwood area per tree (2.9 times) and leaf area per tree (4.2 times), compared with the unthinned stand.

Therefore, although the thinned plantation had not re-equilibrated completely to prethinning conditions at the time of the measurements, the clear increases in leaf and sapwood areas of the remaining trees could be seen as representing a tendency towards complete re-equilibration following a set of homeostatic relationships aimed at maximum site utilization. Lesch and Scott (1997) described an extreme case from South Africa where the rate of growth of young *Eucalyptus grandis* trees under seasonal rainfall conditions was such that any positive effects on streamflow of three rounds of thinning (46, 34 and 50% at age 3, 5 and 8 years) were masked entirely by the steady reduction in flows resulting from the overall vigorous growth of the trees.

Working under wet tropical conditions, Gilmour (1977) was unable to detect any changes in streamflow after 'lightly' logging a rain forest in northern Queensland, Australia. More substantial timber extractions from hill rain forest in Peninsular Malaysia, equivalent to the removal of 33 and 40% of the commercial stocking, affected overall water yield positively, with increases of 40 and 70% respectively.

However, during a 7-year postlogging observation period there was no sign of a decline in the streamflow gain, suggesting that the regrowth in the gaps created by logging remained well below that of the remainder of the forest (Abdul Rahim & Zulkifli Yusop 1994).

This result is contrary to expectation in that soil water levels in gaps, although initially higher than in the surrounding undisturbed forest, tend to decline very rapidly as the regrowth is reestablished. Such findings illustrate the limitations of paired catchment studies if these are not complemented by detailed process studies.

The influence of the configuration of the cutting on the magnitude and duration of any increases in flow has been investigated in some detail in the eastern USA.

The removal of 24% of the basal area from catchment LR2 at Leading Ridge (Pennsylvania) incurred a nearly twofold larger increase in flow than cutting 33% on catchment HB4 at Hubbard Brook (New Hampshire) or catchment FEF2 at Fernow Experimental Forest (West Virginia). The cutting at Leading Ridge consisted of a single block on the lowest portion of the catchment, whereas the cutting at Hubbard Brook took the form of a series of strips situated halfway up the catchment, and that at Fernow Experimental Forest involved harvesting trees from all over the catchment. Therefore, increases in streamflow associated with strip cutting are smaller than for single blocks, which is in line with the idea of increased transpiration by surrounding trees upon opening up of the canopy. No significant differences were found between the cutting *en bloc* of the upper half of a catchment (such as Catchment 7 at Fernow) or the lower half.

Of particular interest to water managers is the role of riparian vegetation which, in contrast to trees growing further upslope, generally has ready access to the groundwater table. Elimination of the riparian vegetation at Coweeta produced a decrease in diurnal fluctuations of baseflow, although the associated increase in water yield was not greater than that associated with the removal of an equal area of forest elsewhere in the catchment.

Apparently, the riparian effect is negligible in areas where soil moisture remains readily available in all parts of the catchment, such as at Coweeta with its well-distributed rainfall. A similar result was obtained for a catchment with deep soils in the summer-rainfall zone of South Africa, but not for an area in the winter-rainfall zone where streamflow gains after cutting pine trees in the riparian zone exceeded those associated with the harvesting of pines away from the stream by 30%.

EFFECT OF FOREST CLEARFELLING ON WATER YIELD

A basic summary of short-term results obtained from paired catchment experiments conducted before 1980 was given by Bosch and Hewlett (1982). Hornbeck *et al.* (1993) and Swank *et al.* (1988) discussed longer-term results, including the effects of multiple treatments, for catchments in the north-eastern and south-eastern USA respectively. Generally, increases in water yield during the first year after treatment are roughly proportional to percentage reductions in stand basal area, provided the latter exceed a threshold of 20-25%. Bosch and Hewlett (1982) made a distinction between coniferous and deciduous hardwood forests, suggesting that increases in flow associated with the cutting of conifers were higher (40mm per 10% change in cover) than for hardwoods (25 mm per 10% change in cover).

However, the scatter around the two tentative regression lines was large, coefficients of determination modest (at 42% and 26% respectively) and no confidence limits were provided. In a later analysis of the same dataset plus an additional 50 studies (mostly from the tropics and Australia), Sahin and Hall (1996) used fuzzy linear regression analysis and obtained the following average changes in flow per 10% change in cover: conifers, 23mm; mixed hardwoods-conifers, 22mm; and hardwoods, 17-19 mm (depending whether annual precipitation exceeded 1500mm or not). Based on this fuzzy regression approach, the overall differences between the respective vegetation groups were therefore smaller than suggested earlier by Bosch and Hewlett (1982). Stednick (1996) approached the problem of the heterogeneity in results for 95 paired catchment studies from the USA by deriving separate regression equations for each of eight geographically homogeneous regions.

In doing so, a clearer picture was obtained of the regional variation in yield increases, which ranged from about 10mm per 10% change in cover in the Rocky Mountains to over 60mm in the Central Plains. Also, some of the regionally applicable regression equations had substantially higher coefficients of determination (*e.g.* 0.65 for the Pacific Northwest or Appalachia, which both had the largest number of experiments), although correlations for other regions were extremely low (*e.g.* 0.01 for the Rocky Mountains). As such, there is something to say for the approach of Trimble *et al.* (1987) who simply pooled the coniferous and mixed hardwood data of Bosch and Hewlett (1982) and added 10 data points of their own, representing reductions in annual flow from large non-experimental ('real world') catchment areas in the Piedmont of the south-eastern USA that had undergone 10-28% reforestation.

The overall coefficient of determination for the combined dataset attained the reasonable value of 0.50 by the inclusion of the Piedmont subset, with 8 of 10 data points from the latter falling within half the standard error of the estimate (compared with only 22 of 55 data points in the Bosch and Hewlett set). This finding is all the more remarkable because half the additional data involved a change in cover of less than 20%, a value below which Bosch and Hewlett (1982) had considered effects on streamflow to be more or less non-detectable (at least on small headwater streams). Apart from contrasts in tree species and age of the vegetation, much of the variability in the change in stream-flow observed for different sites and years after a change in cover relates to differences in climate, catchment exposure and soil depth. The climatic aspect is aptly summarized by the classic statement of Hewlett (1966) that 'it takes water to fetch water'. The importance of catchment exposure is illustrated by the difference in first-year streamflow gain after cutting differently exposed deciduous hardwood forest catchments at Coweeta in the south-eastern USA.

Flows from northerly exposed catchments increased by about 130mmyear^{-1} but increases from southerly exposed catchments could be as high as 410mm year^{-1}. As for the effect of soil depth, Trimble *et al.* (1963), when discussing differences in duration of the changes in streamflow following forest clearing in West Virginia, stated that 'the deeper the soil, the longer it takes roots of new growth or the expanding root systems of residual plants to occupy the soil mantle. When the soil mantle is reoccupied, transpiration reaches maximum levels again, and increases in streamflow disappear'. The role of soil depth can be illustrated further by examining the timing of the maximum and minimum increases in flow after forest clearance during the year. At Coweeta, where precipitation is distributed evenly over the year and the loamy soils are deep (up to 6 m) and capable of storing large amounts of moisture, the soil does not become fully recharged until early spring (April-May).

As a result, the presence or absence of a forest has little effect on the amounts of flow during this time of year (the soils being full of water anyway). However, at the onset of the growing season, increased flow due to reduced water use after clearcutting becomes evident and the contrast in streamflow from forested and cleared areas increases as the growing season advances. The effect continues well into the winter months (November-January), even though the trees have lost their leaves by then and evaporation is strongly reduced. This reflects the development of a much larger soil water deficit below the forest during the summer that takes an additional 2 months (February-March) to be refilled by the winter rains. Conversely, on catchments with shallower or sandier soils (having less water storage and a lower water retention capacity), soil water recharge and depletion occurs much more rapidly and increases in flow after forest clearance follow the pattern of forest water use much more closely.

The effect of the timing of the cutting and the importance of controlling regrowth are demonstrated by a comparison of the results obtained for catchments 2 and 5 at Hubbard Brook (HB 2 and HB 5). On catchment HB 2, the forest was clearcut during the dormant season and regrowth eliminated by herbicide application the following spring. This produced a first-year increase in water yield of 347 mm. In contrast, on catchment HB 5, the removal of the trees took nearly a full year, while regrowth was not suppressed.

The resulting first-year increase in flow thus amounted to only 152 mm, or 44% of that observed for catchment HB 2. In addition, the gain in streamflow had largely disappeared after 3 years. Such contrasting findings once more illustrate the effect of regenerating vegetation on soil water levels and streamflow. However, it should be noted that regeneration was particularly vigorous in the example from Hubbard Brook as it took place mainly via sprouting. As demonstrated by comparative work in West Virginia, elevated streamflow levels lasted much longer (15-20 years) when regeneration had to originate from seeds because of previous herbicide applications than when sprouting was allowed (6-7 years).

Similarly, when a regenerating hardwood forest at Coweeta was cut a second time after 23 years (when streamflow was still about 80mmyear^{-1} above that observed for mature mixed forest), the initial increase in water yield was nearly identical to that of the first cut (375 vs. 362mm). However, the subsequent decline in (extra) water yield was distinctly faster. The difference was attributable to a more rapid recovery of stand biomass during the second regeneration period in association with the greater sprouting potential of the even-aged forest, which contained numerous small stems prior to the second cut. As a result, the (projected) total duration of the second period of increased flows was estimated to be 18 years shorter than that associated with the first cutting (estimated at about 35 years). Such long durations reflect the time required by the roots to reoccupy the very deep soils at Coweeta compared with those of the north-eastern USA where effects are more likely to last 3-9 years.

DIFFERENT PATTERN OF CHANGES IN FOREST WATER

A rather different pattern of changes in forest water use and streamflow with time after clearing has been documented for native mountain ash *(Eucalyptus regnans)* forest in south-eastern Australia. Following clearing and burning, there is an initial rise in streamflow lasting 4-6 years. This is followed by a rapid decrease in flows until the regrowth is about 27 years old, after which run-off totals slowly return to predisturbance levels. The latter may take as long as 150 years. Vertessy *et al.* (1998) provided a mechanistic explanation for this long-term pattern in terms of concurrent changes in overstorey and understorey leaf biomass and transpiration rates, as well as changes in rainfall interception and evaporation from the forest floor.

Results from paired catchment experiments in the humid tropics also show a proportionality in flow increase with percentage of forest removed, as observed under humid temperate conditions. Reported first-year increases after clearfelling range from 125 mm in Nigeria to 820 mm in Peninsular Malaysia and show little relation to amounts of precipitation.

This led Bruijnzeel (1996) to suggest that perhaps the results were determined more by the degree of surface disturbance than by climatic or soil factors. However, in view of the large variability in total evapotranspiration among tropical forests, it may be rewarding to reanalyse the data in the light of contrasts in rainfall interception between regions as well.

EFFECTS OF CONVERTING NATURAL FOREST TO OTHER LAND COVER TYPES ON WATER YIELD

Whilst streamflow totals are observed to eventually return to preclearing levels where regrowth is allowed, the conversion of native forest to other types of vegetation cover may produce permanent changes. For example, permanent increases in annual water yield are associated with the conversion of deciduous or evergreen native forest to agricultural cropping. Reported increases range from 60-125 mm year^{-1} under humid warm temperate conditions to 450mmyear–^{1} in the equatorial tropics. The diminished water use of annual crops compared with that of a full-grown forest reflects the diminished capacity of low vegetation not only to intercept rainfall but also to extract water from deeper soil layers during periods of drought.

The former relates primarily to the lesser aerodynamic roughness of short annual crops (and possibly to their smaller leaf area and thus storage capacity as well), whereas the reduced water uptake of crops reflects their more limited rooting depth. For the same reasons, conversion to pasture generally produces permanent increases in streamflow as well, although the magnitude of the increase depends on precipitation patterns and elevation. Interestingly, Hibbert (1969) noted that replacing deciduous hardwood forest in the south-eastern USA by vigorously growing *Festuca* grass did not lead to increased water yields, although increases of up to 125 mm year^{-1} were observed after the productivity of the grass declined.

Such results can be explained in part by the fact that serious water shortages that could have reduced water uptake by shallow-rooted plants do not develop in the rainy climate at Coweeta, whereas in addition the contrast between forest and grass was lessened further by the deciduous character of the forest. In contrast, the conversion of mixed deciduous hardwoods by eastern white pine *(Pinus strobus)* in the same area brought about a reduction of 250mmyear–^{1} after 25 years.

The decline in flow was ascribed both to a steady increase in intercepted rainfall as the pines grew older and to higher transpiration by the evergreen pine trees during the dormant season.

Similarly, the higher total water use *(ET)* of mature *Pinus radiata* plantations in south-eastern Australia compared with that of native eucalypt forest has been explained in terms of the distinctly higher rainfall interception by the pines. However, where pine plantations replaced non-deciduous native forests elsewhere in the world, streamflows have been reported to return to preconversion levels within 8 years, such as in Kenya and New Zealand.

The results presented thus far pertain mostly to small headwater catchment areas involving a unilateral change in cover. Although these experiments provide a clear and consistent picture of increased water yield after replacing tall vegetation by a shorter one, and *vice versa,* such effects are more difficult to discern in large river basins having a variety of land-use types and temporal changes therein.

In addition, there are the complications associated with strong spatial and temporal variability in rainfall, especially in tropical areas with largely convective rainfall, and the often large-scale withdrawals of water for municipal, agricultural and industrial purposes in populated areas. Qian (1983) was unable to detect any systematic changes in streamflow from catchments ranging in size from 7 to 727km^2 on the island of Hainan, southern China despite a 30% reduction in tall forest cover over three decades. Dyhr-Nielsen (1986) arrived at the same conclusion for the 14 500 km^2 Pasak River basin in northern Thailand, which lost 50% of its tall forest cover between 1955 and 1980.

On the other hand, Madduma Bandara and Kurupuarachchi (1988) observed an increase in averaged annual flow totals for the 1100 km^2 upper Mahaweli catchment in Sri Lanka over the period 1940-80, despite a weak negative trend in rainfall over the same period. Although both trends were not statistically significant at the 95% level, the associated increase in annual run-off ratios was highly significant.

The increased hydrological response was ascribed to the widespread conversion of tea plantations (not forest) to annual cropping and home gardens without appropriate soil conservation measures (Madduma Bandara and Kurupuarachchi 1988).

However, when analysing a longer time series (1940-97) for the 380 km^2 upper Nilwala catchment in Sri Lanka, which experienced a 35% reduction in forest cover during the observation period, a less consistent picture was obtained. Such contrasting findings illustrate the need for high-quality rainfall and streamflow data in historical hydrological data analyses, not only in the context of evaluating the effects of land-use transformations but also those associated with climatic change.

EFFECTS OF FOREST CLEARING ON STREAMFLOW REGIMES

In areas with seasonal rainfall, the distribution of streamflow throughout the year is often of greater importance than the total annual amount *per se*. Reports of greatly diminished streamflows during the dry season after forest clearance abound in the literature, particularly in the tropics. At first sight, this seems to contradict the evidence presented earlier, that forest removal leads to higher water yields, even more so because the bulk of the increase in flow after *experimental* clearing is observed during baseflow or dry-season conditions. However, the controlled conditions imposed during catchment experiments differ from those encountered in many real-world situations.

EXPOSURE OF BARE SOIL AFTER FOREST CLEARANCE

The continued exposure of bare soil after forest clearance to intense rainfall, the compaction of topsoil by machinery or overgrazing, the gradual disappearance of soil faunal activity and increases in the area occupied by impervious surfaces such as roads and settlements all contribute to gradually reduced rainfall infiltration opportunities in cleared areas. As a result, catchment response to rainfall becomes more pronounced and the increases in storm runoff during the rainy season may become so large as to seriously impair the recharging of soil water and groundwater reserves.

When this critical stage is reached, diminished dry-season flow is the result, despite the fact that the removal of the forest has led to reduced evaporation and thus should have induced higher baseflows. On the other hand, if the surface characteristics after forest clearing are maintained sufficiently to allow the continued infiltration of (most of) the rainfall, then the effect of reduced *ET* associated with forest removal will show up as increased dry-season flow. This could be achieved through the establishment of a well-planned and maintained road system plus the careful extraction of timber in the case of logging operations, and by applying appropriate soil conservation measures when clearing for agricultural purposes.

However, even with minimum soil disturbance, there will be local increases in stormflow volumes and peak flows after forest removal because the associated reduction in *ET* causes the soil to be wetter and thus more responsive to rainfall. Relative increases in stormflow response associated with minimum disturbance operations are largest for small events (up to 300%), declining to less than 10% for large events, although the effect will be more pronounced if soil disturbance is severe.

Also, whilst increases in stormflow response after (well-conducted) forestry operations tend to diminish or disappear altogether as the new vegetation cover becomes established, they also may become structural. Examples include run-off contributions from roads or residential areas where there were none before and permanently wetter soil conditions associated with reduced water use. However, it is important to remember that caution is required when trying to extrapolate the undoubtedly adverse effects of indiscriminate forest clearing on local infiltration opportunities and storm run-off generation over large river basins.

HYDROLOGICAL EFFECTS OF (RE)FORESTATION

In response to the widely observed degradation of formerly forested land and the rising demands for industrial wood and fuelwood, the need for large-scale reforestation programmes has been expressed repeatedly (*e.g.* FAO 1986; Postel & Heise 1988). It is of interest therefore to examine to what extent such plantations are capable of restoring the original hydrological conditions, *i.e.* ameliorate peak flows and, above all, enhance low flows. Whilst there is little doubt that annual water yields from forested areas are reduced compared with those for non-forested areas, it must be granted that almost no catchment forestation experiments have been carried out on seriously degraded land.

As such, it could be argued that the clear reductions in total and dry-season flows observed after afforesting natural grasslands and scrublands or pasture in South Africa, New Zealand, south-eastern Australia and Fiji only serve to demonstrate the difference in water use between the two vegetation types. In other words, the potentially beneficial effects on streamflow afforded by improved infiltration and soil water retention capacities during forest development could not become manifest in these experiments. The key question is therefore whether the reductions in storm run-off generating overland flow incurred by such soil physical improvements can be sufficiently large to compensate the extra water use by the new forest, and so (theoretically) boost low flows.

There is no easy answer to this question for several reasons. First, the effect of an increase in topsoil infiltrability on the frequency of occurrence of HOF depends equally on prevailing rainfall intensities.

For instance, rainfall intensities in the Middle Hills of Nepal were generally so low that the 140mmh^{-1} increase in infiltration capacity 12 years after reforesting a degraded pasture site with *Pinus roxburghii* made no difference in the generation of HOF. Similarly, whilst topsoil infiltrabili-ties had roughly doubled over 12 years since reforesting *Imperata* grassland with teak in Sri Lanka, at 30mmh^{-1} the hydrological impact of this increase is probably limited. A much longer time span may be required therefore to restore infiltration capacities to the high values normally found in undisturbed forests.

Secondly, the reductions in catchment response to rainfall observed after forestation (Tennessee Valley Authority 1961; Fahey & Jackson 1997) may well reflect the drier soil conditions prevailing under actively growing forest rather than a reduction in peak-generating HOF. However, none of the studies documenting decreases or increases in stormflows after land-use change has attempted to quantify the associated changes in relative contributions by subsurface and overland flow types. Third, there is also the effect of soil depth, which determines both the maximum amount of water that can be stored in a catchment under optimum infiltration conditions and the possibilities for water uptake by the developing root network of the new trees. And last but not least, there is the confounding effect of rainfall patterns (seasonal vs. well distributed) and general evaporative demand, which both exert a strong influence on tree water use.

Needless to say, the soil and climatic factors, plus the absence of detailed information on the prevailing stormflow mechanisms before and after the conversion, all render comparisons between different sites more complicated.

The only documented real-world case in which the kind of compensation mechanism supposed by Bruijnzeel (1989) seems to have occurred may be the White Hollow catchment in Tennessee, USA (Tennessee Valley Authority 1961).

Prior to improvement of its vegetation cover, two-thirds of this catchment consisted of mixed secondary forest in poor condition (due to fire, heavy logging and grazing), with another 26% under poor scrub. About 40% of the catchment was estimated to be subject to more or less severe erosion. Following extensive physical and vegetative restoration works, peak discharges decreased considerably within 2 years, especially in summer.

However, neither annual water yield nor low flows changed significantly over the next 22 years of forest recovery and reforestation. It was concluded that the extra water needed by the recovering and additionally planted trees was balanced by improved infiltration (Tennessee Valley Authority 1961). However, it is quite possible that the absence of major changes in total and low flows at White Hollow mainly reflects the lack of contrast between tall and short vegetation, as would have been the case when reforesting a truly deforested and degraded catchment.

Support for this contention comes from the observation of Trimble *et al.* (1987) that reductions in flow from several large river basins in the south-eastern USA, which had suffered considerable erosion before being partially reforested, were largest during dry years. In addition, the effect became more pronounced as the trees grew older.

A similar case was recently described for a once seriously degraded catchment in Slovenia, where the spontaneous return of forest produced a steady reduction in annual water yield over a period of 30 years, with the strongest reductions again occurring during the dry summer months.

Therefore, in both these real-world examples the water-use aspect overrides the infiltration aspect, despite the rather modest contrasts in water use between forest and grassland/cropland in the south-eastern USA.

Such findings offer little prospect for the possibility of significantly raising dry-season flows by forestation in the humid tropics. Observed maximum differences between annual water use of fast-growing pines or eucalypts and short vegetation (grass, crops) under well-watered (subtropical conditions attain values of 500-700 mm at the catchment scale and even higher values (> 1100 mm) on individual plots with particularly vigorous tree growth.

The hydrological benefits incurred by the limited increases in topsoil infiltration capacities observed after forestation of degraded land in Nepal and Sri Lanka cited earlier are dwarfed by such high water requirements.

The conclusion that already diminished dry-season flows in degraded tropical areas will decrease even further upon reforestation seems inescapable.

In view of the extent of the low-flow problem, the testing of alternative ways of increasing water retention in tropical catchments without the excessive water use associated with exotic tree plantations should receive high priority.

THE WATER CYCLE

The water cycle, also called the hydrologic cycle, operates similarly to biogeochemical cycles in which plant and animal nutrients move through the atmosphere, the earth, and through living things. The water cycle includes water in the forms of solid, liquid, and gas, and the Sun provides the energy needed to power the constant cycle of water among these three forms. The water cycle occurs in oceans, bays, lakes, ponds, rivers, and streams. The ocean covers most of the Earth's surface and so contributes the largest share to the water cycle; more than 85 percent of the water that evaporates from the Earth's surface into the atmosphere does so from the ocean. About 80 percent of the world's precipitation enters oceans.

The water cycle contains five components that transfer moisture either from the Earth's surface to the atmosphere or from the atmosphere back to Earth. These components are the following:

- condensation—conversion of water vapor into droplets of liquid water
- precipitation—water that falls from the atmosphere to land or to surface waters as rain, snow, sleet, or hail
- infiltration—downward movement of liquid water through soil
- evaporation—conversion of liquid water into gas, or water vapor
- transpiration—movement of liquid water from plant roots, upward in vessels, and into the atmosphere from the leaves as water vapor

Forests serve in the water cycle in two critical ways: transpiration and water storage. Both animals and plants transpire water vapor into the air, but animals transpire as part of aerobic respiration when they exhale moisture from their lungs. Plants and trees possess special cells on the underside of their leaves, called guard cells, that release water vapor from the plant into the atmosphere. The transpiration process begins when trees draw water from the soil through their roots and transport it upward in vessels—xylem carries water and nutrients upward, phloem distributes water and nutrients throughout the plant. Trees, depending on size and species, transpire from

5,000 gallons (18,921 l) to almost 50,000 gallons of water per year, more in warmer weather and less in colder weather. Low relative humidity and increased air movement caused by wind or breezes also increase the transpiration rate. During drought or in the desert where only drought-tolerant plant life lives, a tree's transpiration rate decreases so that it can conserve water. Soil also affects water transpiration. When soil moisture levels are low, as in drought, trees slow their transpiration rate to conserve water. Trees undergoing the normal aging process, called senescence, give way to new trees ready to take their place. Prolonged drought, however, puts all trees into a premature senescence, an event that may eventually kill a forest. Global warming has exerted a critical effect on forest health because it increases the incidence and severity of droughts worldwide, and in doing so it accelerates the death of drought-vulnerable trees.

Forests' second major activity involves the capacity of trees to store water. Trees act as a watershed by absorbing water during floods and storing and slowly releasing water in times of low rainfall. Surface waters, groundwaters (or aquifers), and plant life comprise the Earth's total watershed. Part of trees' role in the watershed involves regulation of the water table, which is the area underground where water has completely filled the spaces between rocks and soil particles. Beneath the water table lies an area of higher density where all the air has been squeezed out to make room for water. This location, called the zone of saturation, holds water undisturbed for longer periods than the water table. Therefore, water moves through three layers in the earth: the upper unsaturated zone, where soils hold varying amounts of moisture (also called the soil zone); the water table, where water exchanges from the saturated layer below to the unsaturated layer above; and the zone of saturation. Tree roots pull water from either the unsaturated zone or the water table.

Human activities interfere with the water cycle in the three following ways: by drawing large amounts from surface sources and groundwaters, by polluting water, and by removing or damaging the world's forests. Part of the problem of water use by the Earth's human population resides in the fact that water is a very poorly managed resource. Water covers 71 percent of the Earth and makes up at least 60 percent of living cells, but only 0.014 percent is available for people's use. Yet humans continue to waste water, pollute it, or otherwise treat it as if it were free. Benjamin Franklin once said, "When the well's dry, we know the worth of water." The world has been progressing toward ever greater water shortages since the acceleration of global warming. Some countries have already entered a dangerous condition known as water stress, meaning their water requirements exceed the water they have available.

Since every living thing needs water and cannot exist without it for more than a few days, maintaining a clean supply is paramount to maintaining biodiversity. Forests play a vital role in maintaining a continuous supply of available water, so this represents one way in which forests maintain biodiversity. The sidebar "The Forest Canopy" point out, forests add to the world's biodiversity in other immeasurable ways.

The Sun's energy powers the water cycle, which is critical for conserving Earth's water. Within this cycle, oceans account for more than 80 percent of evaporation, and more than 80 percent of rainfall lands in the oceans.

Groundwater serves as a major source of drinking water in most of the world, but high water demand due to growing populations draws down the reserves, lowering the water table. Pollution further threatens groundwater reserves. Metals, pesticides, and organic chemicals have infiltrated many once pristine aquifers.

8

Irrigated Forest Plantations

Irrigated forest plantations can be established for the commercial production of fuelwood, posts, construction lumber, and fodder. The use of irrigation practices also allows the use of more exacting fast-growing tree and shrub species. In many instances, the availability of wood from irrigated plantations will lessen the destruction of the natural vegetation.

In arid zones, irrigated forest plantations can be achieved using:

- A dependable and permanent water supply;
- An intermittent water supply;
- Waste water.

IRRIGATION WITH PERMANENT WATER SUPPLY

Depending upon the amount of water which can be made available from a dependable water supply (well, dam, river...), permanent irrigation systems can be established. Different designs of permanent irrigation systems can be chosen, depending on the prevailing conditions. Three types of such systems are reviewed in the following sections: gravity systems, sprinkler systems and localized systems.

GRAVITY SYSTEMS

Gravity irrigation systems are characterized by the manner in which the irrigation stream is controlled by the soil surface. Four types can be distinguished: surface flooding, border check, basin and furrow irrigation.

Surface Flooding

This system resembles the inundation that sometimes takes place on flat lands along rivers and it is the simplest form of permanent irrigation. On gently sloping land that requires little preparation, surface flooding is easy to implement. In essence, water is released from main ditches and allowed to spread over the surface. However simple, this method generally has been inadequate for tree and shrub crops, as it is difficult to obtain uniform distribution of the water. Also, there is a risk that the root system of the plants can become deprived of oxygen because of waterlogging.

Border Check

In this method, parallel earth ridges guide the flowing water as it moves down the slope over the strips which vary from 3 to 30 meters in width and can be more than 100 meters in length.

A relatively large flow of water is needed. The land should have a uniform moderate slope parallel to the checks. Careful land preparation is necessary to ensure stability of the ground. The method is suited to medium-textured, deep, permeable soils. On sandy soils, infiltration would be excessive unless the strips are short. Slow percolation renders it unsuitable for heavy soils.

A drainage ditch should be sited at the end of the strip to carry away any excess water. Design involves achieving an optimum balance between soil type; slope, width, and length of strips; and water flow so that the desired depth of water will be applied uniformly to the compartment without excessive percolation at the input end. In agroforestry applications, it could be ideal because the trees could be grown along the check, the width of which could be sufficient for one or two rows of trees.

Basin Irrigation

This is a system in which the field or compartment is divided into small units each of which has a level surface. The basins are filled with water which is all allowed to infiltrate, any excess being drained off.

When leaching salts from the soil, the depth of water can be maintained for considerable periods by allowing continuous flow into the basins. The method entails relatively high labour inputs.

Furrow Irrigation

This is a common method to distribute irrigation water. Furrows are built from the main feeder channel in parallel lines spaced at regular intervals to permit the wetting of the tree rooting zone.

The width of the furrows and their spacing depend, in large part, upon the permeability of the soil. The heavier the soil, the larger and the wider apart the furrows must be; the opposite applies in more porous soils. The method requires relatively high labour inputs and a high degree of skill and experience in directing water from supply channels into the furrows and controlling its flow.

A disadvantage of the system is that tree roots tend to develop linearly along the furrow; trees tend to lean across the furrows and windthrow sometimes occurs later in the rotation. Unevenness of furrow levels and distribution of water can develop when soil conditions are not uniform. Regular maintenance of the furrows is important.

Sprinkler Systems

Sprinkler systems are most applicable in areas of irregular topography where land grading is not feasible, in irregularly sloped areas, or when rapid application of relatively small quantities of water is desired. Applications to forestry are somewhat limited by crop height and costs, but such systems are applicable in the early establishment phase of forest plantation crops.

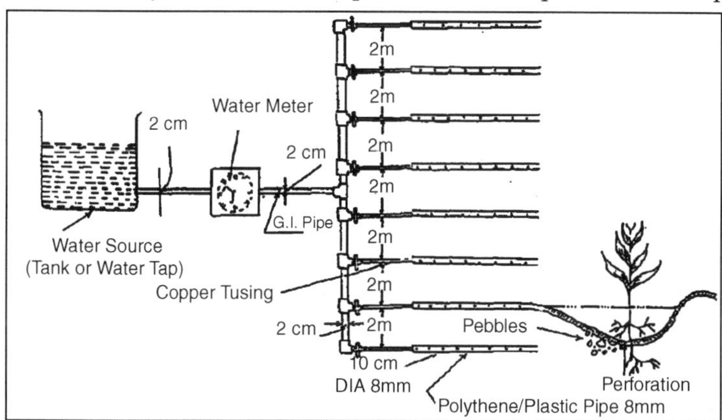

Fig. Drip System of Irrigation

All sprinkler systems have in common a source of water under pressure, a system of pipelines to deliver water to the point of delivery, and nozzles through which the water is distributed.

The advantages of sprinkler systems are that they are applicable to areas of irregular topography and shape without levelling; they can be used in areas of high watertable or where there is a hard pan near the surface without increasing soil salinity; the amount and rate of water application are easily controlled so that run-off and deep percolation can be avoided, this gives them advantages in areas of high permeability. They can utilize a small, continuous supply of water better than gravity methods; water is distributed evenly if it is not too greatly affected by wind; they do not require land for water channels, ditches, and borders, thus saving land and the maintenance costs and inconveniences of open distribution systems.

Localized Systems

Localized irrigation systems, an umbrella term for trickle, drip, drop, or sip irrigation methods, are among those that cause wetting of only part of the soil, *i.e.* that at the base of and surrounding the root system of the plant. They are characterized by slow and low rates of application of water to the plant rooting zone through distribution pipes and orifices or nozzles organized either under or above the soil surface.

The basic components include a pressurized supply of water, a control head, a main line with laterals, and distributors. To create the appropriate pressure in the water supply usually requires a pump and storage tanks or reservoir. The control head is usually sited at the highest point in the field and connected to the water supply. The advantages of localized irrigation systems include the facts that, within limits, they can accommodate undulating ground, are relatively easy to manage, have relatively low labour costs, and are simple to operate. The principal problem of localized irrigation systems is the susceptibility of the smaller pipes and the distributors to clogging by sand, silt, organic matter, algae, bacterial slimes, and precipitation of nutrients, colloidal materials, or lime. Root system size or spread and depth being a function of the volume of water applied at each irrigation, root development can be restricted through inadequate watering. Also, trees can die very rapidly if water is withheld for even a short period: thus the water supply must be reliable.

IRRIGATION WITH AN INTERMITTENT WATER SUPPLY: RAINWATER HARVESTING

Rainwater harvesting as a means of providing water seasonally, over longer periods, has been used in arid areas for thousands of years to grow agricultural crops and trees for fruit, amenity, and other purposes.

Rainwater harvesting essentially involves two components:

1. A catchment or collector area, usually prepared in some manner to improve run-off efficiency, and
2. A smaller water storage area in which crop or tree plants are grown or where water is stored in small tanks or other structures for future use.

In the case of tree planting, rainwater is used directly without storage requirement. Four techniques are particularly relevant:

- Run-off farming.
- Desert strip-farming,
- Contour terrace farming, and
- Flood water spreading.

Run-off Farming

Watersheds are divided into several micro-watersheds depending upon the area needed for each tree. The collection area for each micro-watershed may range from 20 to 1000 square meters, depending upon the area precipitation and tree water requirements. Micro-catchment procedures are used in complex terrain where other water harvesting techniques may be difficult to apply. In a typical situation, a series of well-designed micro-watersheds with appropriate dimensions are prepared. At the lowest point, a basin is dug about 40 centimeters deep and a tree is planted. The depression collects and stores the run-off from the rest of the micro-watershed that feeds the plant. At the root zone, soil should be at least 1.5 meters deep. Diagonal distances between the lowest corner to the farthest contributing corner should be between 5 and 30 meters. This technique is particularly successful in years of normal precipitation. In dry years, most annual crops will fail. It is therefore advisable to select drought-resistant plants for use with this system.

Desert Strip-farming

Although only a very small percentage of the rainfall reaches major stream channels in arid and semi-arid regions, considerable run-off occurs on many of the gently sloping watershed areas. Desert strip-farming makes use of this water by employing a series of terraces that shed water onto a neighbouring strip of productive soil.

Depending upon the topography, soil characteristics, and climatic conditions of the site, two types of micro-watersheds can be used:

1. One-sided micro-watershed for moderately permeable soils and natural land slope greater than 6 per cent, and
2. Two-sided micro-watershed for highly permeable soils with natural land slope less than 4 per cent.

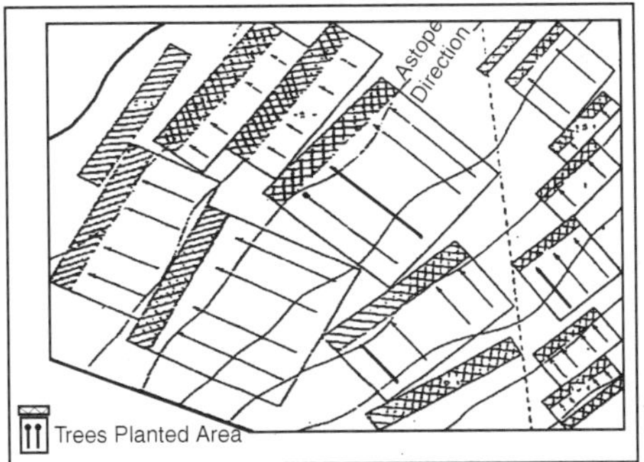

Fig. Run-off Farming Concept

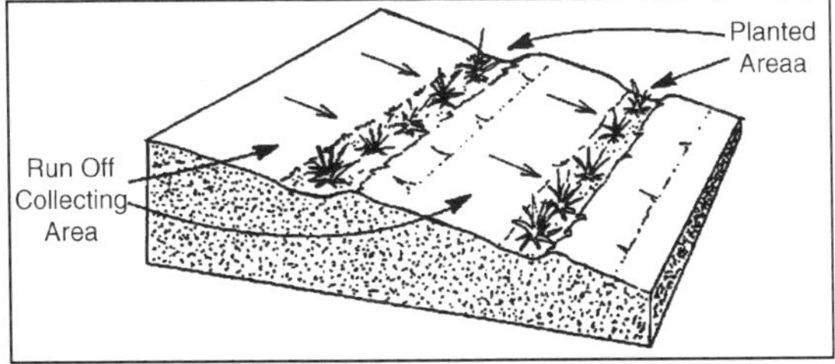

Fig. Desert Strip Farming Concept.

Contour Terrace

The purpose of terracing is to retard and collect all run-off between the terraces. If the run-off is properly managed, enough water can be added to the soil of the terrace to improve tree growth significantly. Terraces are essential on steep slopes where all woody vegetation has been destroyed and is not likely to be reestablished before severe erosion occurs. Terraces should be large enough to hold or carry the heaviest ten-year rain. Contour planting consists of placing long, low barriers perpendicular to the gradients, along contour lines which intercept and retain run-off and silt. The barriers can be of stone, logs, earth or hedge.

Floodwater Spreading

In arid regions, rainfall usually falls during short, intense storms. The water swiftly drains away into washes and gullies and is lost to the region. Sometimes floods occur, often in areas untouched by the storm. Waterspreading is a practice of deliberately diverting the floodwaters from their natural courses and spreading them over adjacent floodplains or detaining them on valley floors. The wet floodplains or valley floors are then used to grow tree or forage crops.

Site selection is the key to success in floodwater farming.

Three principal types of sites are preferred:

1. Slopes below escarpments,
2. Alluvial deltas, or
3. Floodplains.

While potential sites are found in many arid and semi-arid regions, waterspreading systems require careful design and engineering to withstand flood waters. They must be selected so as to optimize topography, soil type, and vegetation.

IRRIGATION WITH WASTE WATER

The use of waste water for agricultural crops and tree production constitutes an interesting approach for many arid countries.

It can be justified by the following facts:

- Creation of a resource from waste water. Fertilizer equivalent contained in waste water from an urban centre averaging 10,000 inhabitants without polluting industry amounts to some US$30,000.
- The technique can be considered an economical method for waste water treatment.
- It is a water saving device so essential for dry regions where water resources are scarce.

Fig. Sketch of a water spreading system.

Several trials have shown that irrigation with waste water improve tree growth. However, a number of facts remain to be documented, such as the optimum size of equipment, the most adequate techniques and the resolution of health problems associated with some techniques in practice.

Depending upon the available water resource and the type of exploitation, three situations can be distinguished:

- Untreated waste water,
- Partially treated waste water, and
- Completely treated waste water.

Untreated Waste Water

The use of untreated waste water is usually impracticable on account of the odour. Untreated waste water is either transported by tankers from houses and factories and poured into prepared trenches, or spread directly in the area.

Odour could be rendered less obnoxious by mixing untreated waste water with urea or ammonium sulphate and muriate of potash. There are two problems associated with this situation. The first is the organization of waste water transport, and the second, the agreement between the producer and the user of waste water as to when the water is needed by the user.

Partially Treated Waste Water

This technique aims at reducing the cost of water treatment by replacing the water softening plant with a simple primary water softening device. Soil will complement the full treatment This technique is also used to recuperate fertilizer contained in the semi-treated water.

Completely Treated Waste Water

This situation is very common in many arid countries. Urban centers and nearby plants treat their water. Agricultural farms nearby can use this water by connecting water pipes to the main treated water canal. This treated water can be totally used for growing agricultural crops and forest trees, partly reused by the plant or poured into a nearby canal.

Technical Considerations for Irrigated Plantations Using Waste Water

The use of waste water for irrigated forest plantations requires a periodical chemical water analysis because the quality of waste water varies considerably. In addition to water pH, consideration should be given to the variation in the amount of chloride, sulfate, free ammonia, water fluorine, phosphate, zinc, boron and silica. Soils for plantations should have a good porosity without a high infiltration rate. Soil absorption capacity should be monitored to follow its evolution.

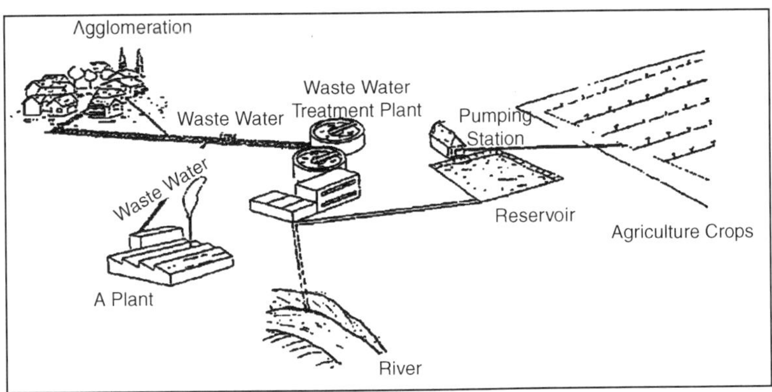

Fig. Partially Treated Waste Water

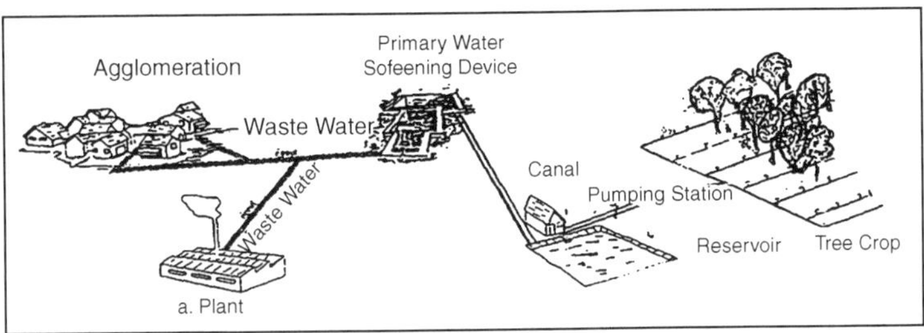

Fig. Completely Treated Waster

Health aspects

There are a number of health hazards associated with the use of waste water. Contamination of soil and vegetation, effluent run-off polluting water canals, waste water infiltration to groundwater, germ dispersal through biotic and wind factor.

This necessitates limiting the use of waste water when untreated or partially treated to sites with low slope, and with moderately permeable soil. Sites should also be distant from human settlements. Irrigation techniques by gravity limit the direct contact of waste water with the plant but have the disadvantage of contaminating underground water if waste water is applied in great quantity. Sprinkler irrigation techniques limit the risk of underground water pollution but lead to direct contact of waste water with plants. Localized irrigation systems can prove the most apt techniques provided the problems of clogging of hoses are solved.

Total security can be ensured only through a set of preventive measures adapted to each site:

- Selection of suitable tree crop (forest tree species for wood and fuelwood are the least affected);
- Selection of appropriate site and irrigation techniques;
- Addition of appropriate decontamination products to waste water.

TECHNIQUES OF FOREST PLANTATIONS

To appreciate the need for forest plantations in arid zones, the roles played by these plantations must be defined.

Quite often, there are a number of roles (such as fuelwood or fodder production) which, through careful planning, can be combined to achieve multiple benefits.

The manual describes the techniques for the establishment and management of forest plantations in arid zones.

SITE RECONNAISSANCE

The more information there is available about the site conditions in the area being considered for tree and shrub planting, the better are the chances of selecting the tree and shrub species best suited to the area.

Information most commonly included in site reconnaissance is:

- *Climate*: Temperature, rainfall (amount and distribution), relative humidity, and wind.
- *Soil*: Depth of soil and its capacity to retain moisture, texture, structure, parent material, pH, degree of compaction, and drainage.
- *Topography*: Important for its modifying effects on both climate and soil.
- *Vegetation*: Composition and ecological characteristics of natural and (when present) introduced vegetation. On areas which have not been degraded by man, the vegetation can provide an indication of the site. Unfortunately, over much of the arid world, the vegetation has been so disturbed that it is no longer a reliable indicator of potential planting sites; in these situations, site selection should be based on soil surveys.
- *Other biotic factors*: Past history and present land use influences on the site, including fire, domestic livestock and wild animals, insects and diseases.
- *Watertable levels*: A knowledge of the depth and variation of the watertable levels in the wet and dry seasons is valuable and can be crucial in determining the tree and shrub species that can be grown. Watertable levels can be estimated from observations in wells or by borings made for this purpose.
- *Availability of supplementary water sources*: Ponds, lakes, streams, and other water sources.
- Distance from nursery.
- Apart from the above biophysical information, socio-economic factors also play an important role. Among these factors are:
- The availability of labour.
- Motivation of the local population.

- The distance of the forest plantation to the market and consumer centers.
- Land ownership and tenure.

SELECTION OF THE PLANTING SITE

Where to plant is generally a collective decision made by policy makers, foresters, and the planting crews, based on information obtained in the site reconnaissance. The key is to select the site that, when planted, will lead to the establishment of a successful forest plantation. Often, the choice of the planting site is limited to lands which are not suited for agriculture or livestock production; when this is the case, the site reconnaissance information gains importance. The boundaries of the planting site, once the area has been chosen, should be marked with boundary posts. When there is a danger of trespassing and damage by grazing animals, a boundary fence should be established. Fencing is costly and, therefore, should only be built when other means of protection are not effective. Once a forest plantation is well established and the trees are sufficiently tall, the fences can be removed and reused at another planting site. When roads and other passageways traverse the planting site, they should also be contained with fences. In many instances, tree and shrub planting is undertaken to protect fragile sites from degradation. However, in some situations, the fragile sites should not be planted; it may be better not to disturb the soil in these areas. Where gullies have been severely degraded by erosion, protective measures other than the planting of vegetation (such as building small checkdams) may be necessary.

SPECIES SELECTION

When the best possible information has been collected on the characteristics of the site to be planted, the next step is the selection of the tree or shrub species to plant. The aim is to choose species which are suited to the site, will remain healthy throughout the anticipated rotation, will produce acceptable growth and yield, and will meet the objectives of the plantation (fuelwood production, protection, etc.).

For a successful planting, performance data may have to be extrapolated from one locality to another. Results from a locality where a tree or shrub species is growing (either naturally or as an exotic) strictly apply only to that locality; their application in another locality involves the assumption of site comparability, an assumption which may or may not be justified. When reliable information shows a close similarity between the site to be planted and that on which the species is already successful, it is generally possible to proceed to large-scale planting with confidence.

In practice, the above data are seldom available, and planting on the new site becomes (in effect) experimental and should proceed on a small scale; when this occurs, detailed performance records should be maintained throughout the experimental planting period. The selection of tree or shrub species through the use of analogous climates is important as a first step; but this must be amplified by an evaluation of localized factors which can be more important (for example, soil, slope, and biotic factors). However, the ability to match closely a planting site and a natural habitat may not preclude the need for species trials, since climatological or ecological matching may not reveal the adaptability of a species. It cannot be emphasized too strongly that, without such trials, the choice of tree or shrub species is (in most cases) a risky business. Since planting in arid environments is normally an expensive undertaking, large-scale failures which result from the wrong choice of species or failure to test them can prove costly.

PREPARATION OF THE PLANTING SITE

When the tree or shrub seedlings arrive from the nursery, the site should have been prepared to ensure that planting can proceed without delay. Arid zone conditions frequently demand more intensive and thorough site preparation than is necessary for planting programmes in moister climates.

Objectives of Site Preparation

Among the objectives of site preparation in arid zones are to:

- Remove competing vegetation from the site.
- Create conditions that will enable the soil to catch and absorb as much rainfall as possible. Surface run-off should be reduced to increase the moisture in the soil.
- Provide good rooting conditions for the planting, including a sufficient volume of rootable soil. Hardpans must be eliminated.
- Create conditions where danger from fire and pests is minimized.

Site preparation is directed towards giving the seedlings a good start with rapid early growth. In general, the methods used to achieve site preparation will vary with the type of vegetation, amount and distribution of rainfall, presence or absence of impermeable layers in the soil, the need for protection from desiccating winds, and scale of the planting operations. Additionally, the value of the tree or shrub crop to be grown is important in determining the amount of expense that may be justified in plantation establishment.

Methods of Site Preparation

In general, preparation of the site by hand is possible and economical only for relatively small-scale projects, where the labour of clearing the competing vegetation and working the soil is not too time-consuming. Under certain conditions, animal-drawn ploughs and harrows can also be economical for small-scale operations. Mechanical soil preparation, used increasingly in large-scale planting programmes, has become a common practice in many areas; often, this is because the supply of labour and the time available for ground preparation are too limited to permit large-scale projects to be undertaken by hand. Some operations, such as deep subsoiling and the breaking up of hardpans, can only be done by machines.

Whatever method of site preparation is used, a planting pit (of an appropriate size) should be prepared. The objective of creating planting pits is to aerate and loosen the soil in which the plants will grow. When these planting pits are prepared, they should not be left empty with the excavated soil lying on the ground, but refilled immediately, otherwise sun and wind will dry out the soil completely.

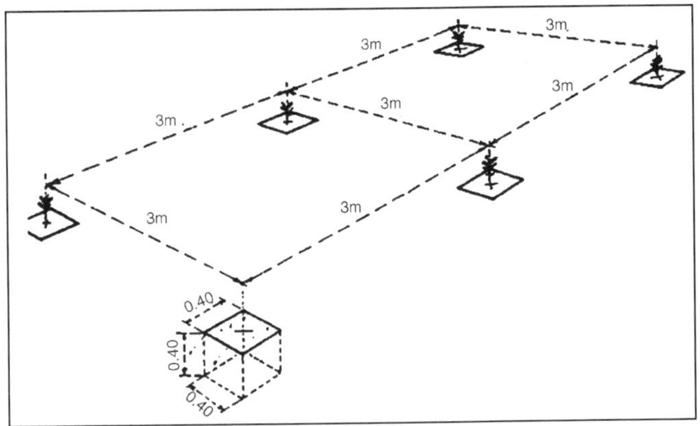

Fig. Planting Holes 0.4m × 0.4m × 0.4m at a density of 3m × 3m.

Fig. 1B One Year Old Eucalyptus Planted in Holes (Gujaab) India

Soil preparation can be carried out in patches, strips, or by complete cultivation. Complete cultivation is necessary for tree and shrub species which are intolerant of competition from grass, forte, and woody growth (such as most eucalyptus species). Sometimes, spot preparation may be sufficient, but the spots should be large (for example, 1 to 1.5 meters in diameter). Also, it is important that the work be done thoroughly. Other methods of soil preparation by hand are the ash-bed method, tie-ridging, contour trenching and terracing, and the "steppe" method.

The ash-bed technique consists of piling the debris from harvesting or clearing the land into long lines or stacks. After drying, the debris is burned and vegetation is planted in the ash patches. Sometimes, the lines or stacks of debris are covered with "clods" to obtain a more intense heat when burning.

Advantages of this method are that the burning kills the competing vegetation, the area remains free of this vegetation for an appreciable period, and the ash provides a useful fertilizer for the planted trees or shrubs.

The tie-ridging technique involves the cultivation of the entire area and the establishment of ridges at specified intervals. The main ridges, aligned along the contours, are joined by smaller ridges at right-angles to create a series of more-or-less square basins which retain rainwater and prevent erosion. The ridges are generally 3 meters apart. The trees and shrubs are planted on the ridges. This method is suitable for flat or gently sloping ground and can be combined with an agricultural crop during the initial years of plantation establishment. Trenching techniques along the contours are used in site preparation in hilly country. The trenches can be continuous, divided by cross banks, or consist of short discontinuous lengths, arranged so that the gaps between the trenches in one row are opposite those in the next row; in this latter instance, run-off from rainfall is caught. Trenches are formed manually or mechanically. On gently sloping ground, the herring-bone technique can be used).

Terraces, which are wider and flatter than trenches, can be either manually or mechanically formed on the side of a hill by digging soil from the uphill side and depositing it on the downhill side. Usually, the bottom of the terrace is made to slope into the hillside. The purpose of terracing is to retard and collect water run-off between the terraces. Because of the improved soil moisture conditions, the terrace provides improved conditions for plant growth. Planting is done on the ridge of soil, at the base of the ridge, or in patches at the bottom of the trench, according to moisture conditions. Terraces are used widely on moderate to severe slopes. Terraces can be 2 to 3 meters or several hundred meters in length. If short, they can be staggered on the hillside wherever convenient. Sometimes, crescent-shaped terraces are constructed with the two tips of the crescent pointing uphill.

The "steppe method of site preparation is designed to promote growth of trees and shrubs in extremely dry areas. In this method, the surface of the soil is modified by breaking-up and stirring the deep layers of the soil with rooters, rippers, or large discs, and then building widely-spaced, parallel ridges following the contour. Ridges are built with the topsoil, and trees or shrubs are planted on the lower half of the ridges facing the slope; here, the depth of moist soil is greatest, due to accumulation of water after rain. The purpose of the "steppe" method is to maintain a reserve of moisture in the deep layers of the soil. Spacing between ridges is greater with lower rainfall, as the catchment area between the ridges is increased.

TIME OF PLANTING

The planting season generally coincides with the rainy season; usually, planting is started as soon as a specified quantity of rain has fallen. This amount of precipitation must be judged on the basis of local knowledge. Planting can also be initiated when the soil is wet to a specified depth (approximately 20 centimeters).

A common mistake is to start planting too soon. On the other hand, if planting is started too late, it may be difficult to complete a large planting programme in the scheduled time, and the plants will lose the maximum benefit of rains after planting; this can be a serious matter where the rainfall is low and erratic.

PLANTING OF CONTAINERIZED STOCK

Planting of containerized stock is usually done in holes that are large enough to take the containers or the root-balls when the plants are removed from the containers. It is essential that the surrounding soil is firmed down around the plant immediately after planting to avoid the formation of air gaps which can lead to root desiccation.A good practice for the preparation of planting holes is to surround the planting pit with a small ridge (15 to 20 centimeters in height) of soil, to obtain a small basin (about 80 centimeters in diameter); this is especially helpful when the plants are watered individually after planting. The small prepared basin can also be covered with a plastic sheet (held in place on the ground with stones or earth), with an opening in the centre for the plant. The plastic sheet impedes evaporation of ground water from the planting hole; also, dew collects on its surface and runs to the central opening of the sheet to irrigate the roots. Through conservation of soil moisture, plastic films facilitate more rapid establishment and growth of trees and shrubs during the initial, and most critical years. Another benefit of opaque plastic films is that they inhibit weed growth by reducing light penetration. With the suppression of weeds in the immediate vicinity of the plants, labour also can be saved.

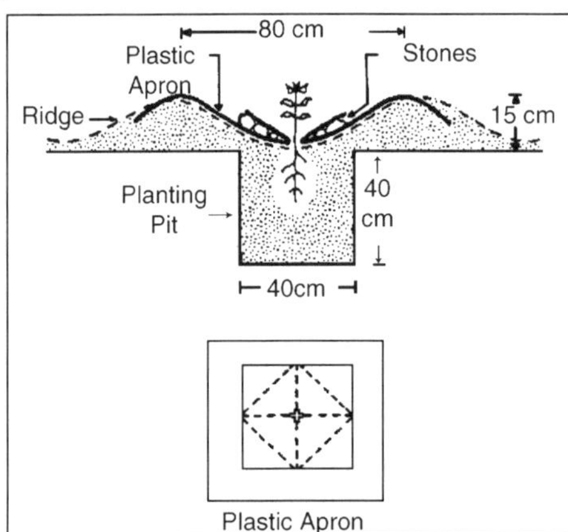

Fig. A Planting Hole with Plastic Apron toImpede Evaporation of Ground Water.

A threat to newly-planted trees in arid zones is the high rate of transpiration. Unless the plants can establish themselves quickly and compensate for the transpiration by taking water through their root systems, they will wilt soon after planting. This explains why even a single watering immediately after planting can be useful. In general, containerized seedlings have a distinct advantage over barerooted seedlings, in that the earthball surrounding the roots provides protection during transport and enables the plant to establish itself quickly and easily.

The restriction of lateral root extension, a result of using containers, can cause root malformation, coiling, and spiralling. In extreme cases, the coiling can lead to strangulation of the roots and the death of the plant. In other situations, it may reduce wind-firmness or lead to stunted growth. Unfortunately, the symptoms may not become apparent until 4 to 5 years after planting. To reduce the damage of root malformation in containerized plants, a common practice is to remove the container from the soil cylinder before planting and make two or three vertical incisions to a depth of one centimeter with a knife to cut "strangler" roots. As a further precaution, the bottom 0.5 to 1 centimeter of the soil cylinder can be sliced off. Care must be used to ensure that the soil does not disintegrate and expose the roots to desiccation.

THE TIMESCALE OF FORESTRY

In many cases, attachment to a particular technique of food production is reinforced by considerations derived from the timescale of forestry. Historically, many rural populations have developed a dependence upon outputs of the forest, because the latter existed as an abundant available local natural resource which could be utilized almost at will. As long as the wood products remained abundant, this process of exploitation could take place without regard to the relatively long time involved in producing wood of usable sizes. However, once the point is reached where wood can be supplied only by growing it, the timeframe involved can become an important limiting factor. Frequently, the timescale of forestry is in conflict with the priorities of the rural people; logically, these priorities are focused on meeting basic present needs. Present needs are imperative, particularly in subsistence situations. Land, labour, and other resources which could be devoted to providing the food, fuel, and income needed today cannot easily be diverted to the production of wood, which will be available only several or many years into the future.

Forestry can continue to exist or be introduced only if it allows the present needs of the people involved to be met. If local tree and shrub cover still exists, it may be possible to provide the same level of production in a less destructive way. In some areas, for example, destructive local cutting of the forest may be halted and reversed by concentrating the cut in designated areas and at designated times, and protecting the rest of the area so that natural regeneration can occur. In other areas, forestry may be introduced in conjunction with other activities which secure income to meet the needs of the farmer until his trees yield products.

The Spatial Distribution of Benefits

Considerations of the spatial distribution of forest benefits are important. To the farmer, the forest is land upon which to cultivate his food and cash crops, a source of fuel and building materials, and a source of fodder and shade. The fact that the trees and shrubs that are utilized could provide the raw material for an economically-important industry elsewhere is of little relevance to him. To expect the farmer to adapt his way of life to accommodate the interests of others is unrealistic. Therefore, the build-up of more stable forest crop/tree systems will only occur when the community, in some way, benefits in appropriate measure from any change in the pattern of natural resources use.

The core of the problem for rural people is usually that they derive insufficient benefits from the forest; in general, this is attributable to conventional forest management objectives and administrative practices, such as an orientation of natural resources towards conservation, wood production, revenue collection, and regulation through punitive legislation and regulation. Consequently, the task of a forester is to engage the people more fully, positively, and beneficially in management, utilization, and protection; this may take the form of greater participation by the people in forest work, development of the income potential of miscellaneous products that can be produced in the forest, or the allocation of forest land for the concurrent production of wood and agricultural crops, or of forestry and livestock.

The issue of distribution of benefits can also arise with management designed to establish industrial wood crops through farming systems which grow trees or shrubs with food and cash crops. In themselves, the trees and shrubs may bring no direct benefit to the farmer; rather, they are an impediment which complicates his task. These systems will succeed only when the farmer perceives adequate compensation. Land, in itself, is not a sufficient inducement, other than perhaps in the short-term; it has been observed that, over time, multiple-use systems tend to evolve either into settled agriculture (with the rejection of the associated growing of trees or shrubs) or into full-time forestry employment.

Institutional and Technical Constraints

There remain situations in which there is no lack of interest in forestry nor any conflict with other aspects of the way of life, but only a lack of organization or means. Areas which are marginal for agriculture may also be marginal for forestry; this is particularly the case in arid areas, which tend to impose severe climatic constraints on the growing of trees and shrubs. Additionally, arid environments impose other constraints, including that of availability of labour. In arid areas, the planting seasons for both agriculture and forestry are short and coincide. As a result, the availability of labour for tree or shrub planting can be restricted, and planning must allow sufficient flexibility to overcome such a constraint. Another constraint to forestry in arid environments is the associated cost. Successful tree or shrub planting programmes often involve elaborate site preparation and planting techniques which require sophisticated" and costly equipment. It may be an activity which is beyond the capability and resources of the local community. Therefore forestry, as an activity which the community can implement, is often confined to the manipulation of existing vegetation, with plantation forestry undertaken by the responsible technical branches of government.

Land management on steep upland areas is also likely to present problems of soil stabilization or control of water run-off; establishment of forest cover on parts of the watershed must be accompanied by conservation measures (such as the construction of terraces) to permit stable crop production on other parts. In many instances, farmers will not have the resources to do this. To establish terraces, for example, they would have to forego one crop. Therefore they will need external support, such as credit and food aid.

Another institutional issue is security of land tenure. Patterns and traditions of use on tribal or communal land often make no provision for usages such as forestry, which require the setting aside of land for a particular purpose for relatively long periods of time. Farmers must have assurance that they will control the land on which trees or shrubs are planted.

Frequently, the lack of a tradition in forestry extends beyond just a lack of knowledge about growing trees and shrubs or of an appropriate institutional framework within which to implement it. Usually, it contrasts with a "deep" tradition in agriculture. This contrast is reflected in the attitudes the people have towards forestry, which are significantly different from attitudes towards agricultural crops and livestock. A forest often tends to be seen as a negative element of the environment by farmers. To the settler, it is an impediment to the clearing of his lands and a haven for his enemies.

These views can persist long after the forest has receded from the immediate vicinity of a farm or community. Hostility towards forests and trees can persist in areas which already experience shortages of fuelwood and building poles, because of the damage done to agricultural crops by birds which roost in trees. Other institutional constraints can also limit the success of community forestry approaches: poor seedling distribution systems, poor management, lack of village funds in order to pay for labour, lack of labour at the right time, planting ill-suited exotics vis-a-vis other options for natural stand management, and so on.

Providing protection to communal woodlots is a particularly frequent problem which may sometimes be reduced by staggering tree planting over several years so that the area which must be protected at any time is kept as small as possible. When the trees have reached a stage where they will be safe from livestock damage, the area can be opened for grazing and browsing. People might also be given access to protected areas for fodder collection, and it is quite likely that fodder production will be greater on these lands if they are protected from free grazing.

The type of silvicultural management system which is adopted is also important. For example, if the community's on-going needs are to be met, annual cutting cycles may be required. However, if the woodlot is very small and the size of the community large, it may not be feasible to divide a small harvest on an annual basis. Similarly, if local people are to be involved in the harvesting, socially workable schemes such as clear-cutting may be preferable to more technically difficult systems like selective thinning. Any management system must, of course, be within the technical competence of the user group and be simple enough that members will feel confident that they can control it in practice.

THE FOREST STATUTE

The Forest Statute provides for the conservation, protection and efficient management of forests taking to account their ecological, social and economic significance, and basing decisions on available scientific information. The aims of this regulation are to conserve natural forest features, increase regeneration and productivity of forest areas, enable efficient use of forest resources, protect rights of use with respect to forests, and to strengthen the legal enforcement of forest protection. However, the Forest Statute does not take account of the implications of land privatization, and still recognizes all forests as absolute property of the State. This restricts local authorities and the private sector from taking over some of the responsibilities for forest protection, and increasing forest cover. At present the Ministry of Nature Protection is involved in drafting amendments to the Forest Statute, in order to bring it up to date.

Since all forests are considered as State property, the government is responsible for overseeing the protection and management of forests, through relevant local authorities ('marz') and a special State agency ('Hayantar', under the Ministry of Nature Protection).

The Government is responsible for: determining priorities for management, classifications of forests, procedures for conservation and management, and forest use fees; approving appropriate forestry projects; setting quotas for timber extraction; implementing research and forest management practices; monitoring of forest stocks; and collaboration with international authorities relating to forestry practices and conservation.

Local authorities are responsible for: provision of temporary concessions and supervision of construction, industrial and mining activities in forest areas.

The authorized State agencies ('Hayantar') are responsible for: the development and implementation of forestry projects; registry of forest stock; ensuring forest regeneration, management and sustainable use; and for addressing issues linked to mismanagement of abuse of forest resources, including illegal felling and the effects of pollutants. A range of other provisions are made within the Forest Statute, including: age of maturity and felling; forest use; rights and obligations of forest-users; methods for timber extraction and processing; use of forests for research, hunting and recreation; measures for forestry in protected areas, and urban regions; rates of forest regeneration; charges for forestry use; supervision of forestry; measures for resolving disputes and reactions to violation of forest legislation; and development of international agreements relating to forestry.

THE LAW ON SPECIALLY PROTECTED AREAS

The Law on Specially Protected Areas outlines the procedures for establishing protected areas and guides their management. The aims are as follows: to maintain the balance of natural ecosystems, to preserve natural monuments of national importance, to conserve the biodiversity of the country, to control use of natural habitats, to promote environmental education and public awareness and to ensure recognition of natural resource depletion within the legal framework. The law specifies that protected areas be established trough government decree, and that overall responsibility for their management lies with the Ministry of Nature Protection. The law also refers specifically to the development of a State listing for protected areas, mechanisms for protected area identification and gazettement, and the status and management regimes for different types of protected area. However, this law does not address a number of relevant issues, such as socio-economic benefits of biodiversity, land privatization, and the role and rights of the private sector, and might therefore be improved by revision.

DRAFT LAWS ON FLORA AND FAUNA

The draft laws on Flora and Fauna set out policies for the conservation, protection, regeneration and management of natural populations of plants and animals, and regulations for human impacts on natural diversity. The Law on Flora aims to ensure sustainable conservation of plants, their genetic diversity and natural habitats, to develop scientific assessments of levels for sustainable use of natural plant populations, to ensure a sustainable conservation of flora, and to protect the rights of those involved in plant conservation and management. The implementation of this law will be overseen by the Ministry of Nature Protection, and by local government and other agencies.

The draft law provides for: inventory, study and monitoring of plant populations; development of a State listing for plants and their use; further elaboration of the Red Data Book for plants; investigation of issues relating to plant conservation; conservation of rare and threatened plant species; use of plants; rights and obligations of plant collectors; restriction or termination of rights to collect particular plants; measures for dealing with disputes over use of plants; and international agreements relating to plant conservation issues.The Law on Fauna aims to: ensure conservation of animals and their genetic diversity, maintain the integrity of animal populations, protect animals from inappropriate disturbance, protect migration routes and regulate use of animal species.

The responsibilities of different agencies (including the government, ministries and other State bodies, local authorities and local self-government institutions) are outlined. The draft law make provision for: survey, study and monitoring of animals; listings of animals and their use; elaboration of the Red Data Book for animals; setting goals for animal conservation; measures for dealing with disputes; and international agreements relating to animal conservation issues.

STATE STRUCTURES FOR BIODIVERSITY MANAGEMENT

Since 1991 the Ministry of Nature Protection and Ministry of Agriculture have been responsible for the management of natural resources in Armenia.

The Ministry of Nature Protection is responsible for ecological survey and inventory, monitoring and management of biodiversity (both in-situ and ex-situ), development of guidelines for sustainable use, and for reviewing the implementation of legislation relating to the environment.

The Ministry has responsibility for a number of protected areas, and also oversees and supervises activities of other government agencies in relation to natural resource use. In line with current legislation, the Ministry of Nature Protection considers all applications for the development of new industrial enterprises, and a statutory environmental impact assessment enables the Ministry to offer advise, monitor and, where necessary, block such developments. The Ministry also issues licences for hunting and collection of medicinal plants, and there are plans to extend the licensing system for other forms of natural resource use, involving further regulations relating to dates, appropriate collection methods and fees.

The Ministry of Agriculture is responsible for management of state agricultural lands, and for supporting farmers of privatized land. In addition, the Ministry of Agriculture oversees management of agrobiodiversity, and manages six of the State Reservations in Armenia.

A number of other state agencies are involved in the management of agrobiodiversity, under the supervision of the Ministry of Agriculture. These include: the State Department of Geology, the State Land Inspectorate, the State Veterinary and Animal-breeding Inspectorate, the Department of Livestock Breeding, and the Department of Plant cultivation, Selection and Land management. Regional rural inspectorates have some devolved responsibilities for biodiversity management and supervision. These inspectorates include the State rural inspector for animal-breeding, the State rural inspector for land-use, and the State rural inspector for geology.

SCIENTIFIC RESEARCH

Armenia has a strong tradition of scientific research relating to biodiversity and other fields, and this has been maintained despite recent economic crises. Research has focused on inventories of flora and fauna (and their genetic diversity), taxonomy, population dynamics, adaptation by exotic species, plant phenology, identification of impacts on biodiversity and studies of effective conservation measures. A range of collections provide the basis for much of this research (including collections based in herbaria, museums, botanical gardens, as well as seed-banks, and genetic databases).

For example, the Institute of Botany (of the National Academy of Sciences) holds more than 265,000 plant specimens, while a further 25,000 are held in Yerevan State University. The largest collections of animals are based in the Institute of Zoology and in the Natural History Museum of Armenia, with a total of 10,000 specimens, including many rare and threatened species (such as Sevan trout, leopard, manul, wild cat, striped hyena and Armenian mouflon). A permanent research station and pilot plot (40 ha.) has been maintained at altitude of 3200m on Mount Aragats since 1962. Extensive research on alpine flora and fauna has been conducted at this site, including patterns of evolution, and assessing human impacts on alpine vegetation, with the aim of developing mechanisms for effective conservation of such habitats.

A number of institutions are involved in scientific research relating to biodiversity, including: The National Academy of Science (comprising the Institutes of Botany, Zoology, Bacteriology, Aquatic Ecology and Fisheries, Agro-chemistry and Hydroponics, and the Centre of Noosphere Research); Institutes of the Ministry of Agriculture and Ministry of Industry (Institutes of Land Cultivation, Soil Science, Fruit and Grape Cultivation, and Applied Bio-technologies); The university sector (including the Yerevan State University, Agricultural Academy, Medical University, and Teaching Institute).

BIODIVERSITY INVENTORY AND MONITORING

Biodiversity monitoring is an important component of any national ecological monitoring programme. Systematic monitoring of biodiversity is not currently conducted in Armenia, and at present, monitoring is not even conducted in protected areas, as a result of lack of resources and of qualified staff to undertake systematic surveys.

Although inventory work on the habitats of Armenia was initiated in the 1920s, a full inventory of flora and fauna has never been conducted, and much of the current information is out of date. Between the 1950s and 1980s floral inventory work focused on the region around Yerevan. In addition, some rare species (including big mammals) and game species were periodically surveyed during this time. Inventories of vascular plants, fish, birds and mammals have been conducted in a number of protected areas (Erebuni, Dilijan, Khosrov, Sev Litch, and Lake Sevan National Park).

Although in many cases these are out-dated or incomplete, such surveys may provide important baseline data for future monitoring programmes. Furthermore, much of the research (including phenological studies) conducted by the National Academy of Sciences and other institutions has important value for the development of appropriate monitoring methods, as well as providing valuable baseline data. For example, the information gathered on Armenian plants during collections for Yerevan Botanic Garden could also be related to distributions and trends in populations.

An improved system for monitoring biodiversity might involve a range of levels: surveys of landscapes, ecosystems and species, assessments of human impacts, and prediction of likely consequences. The development of any monitoring system should take into account existing studies of biodiversity, current protection status, and the aims and indicators of monitoring. An effective monitoring system will require country-wide co-ordination, permanent survey sites, and availability of the necessary technical abilities, staffing, equipment and communication systems.

Armenia has ratified a number of international agreements and conventions relating to the protection of biodiversity.

- Convention on Wetlands of International Importance Especially as Waterfowl Habitat. Armenia ratified the Ramsar Convention in 1993, however despite the international importance of Lake Sevan and Lake Arpa, little has been done to implement this convention.
- Convention on Biological Diversity. This convention was ratified by Armenia in 1993, and the first stage of implementation is currently being undertaken including the development of a National Biodiversity Strategy and Action Plan, and this first National Report (incorporating a Country Study of Biodiversity) to meet reporting requirements to the convention.
- Convention concerning the Protection of the World Cultural and Natural Heritage (World Heritage Convention, Paris, 1972). This convention was ratified in 1993, however there is little available information on implementation.
- Convention to Combat Desertification. The UNCCD was ratified by Armenia in 1997. A project is currently being developed to meet obligations under this convention.
- Framework Convention on Climate Change (UNFCCC, Rio de Janeiro, 1992). The UNFCCC was ratified by Armenia in 1993, and production of a Country Study on Climate Change is underway.
- Three further international conventions relate to biodiversity conservation, and Armenia has not yet acceded to these.
- Convention on International Trade in Endangered Species of Wild Fauna and Flora.
- Convention on the Conservation of Migratory species of Wild Animals.

- Convention on the Conservation of European Wildlife and Natural Habitats.

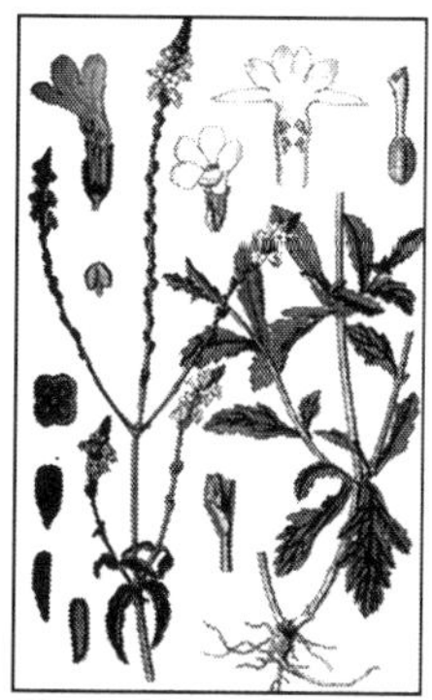

Everything in Europe is about to change due to the EU Food Supplements Directive (FSD), which will soon go into full effect. As a result, around 5,000 safe formulas and nutrients that have been on the market for decades will be banned in fifteen European countries. This directive regulates that if a plant or herb is medicinal or effects physiological function in any way, it is a medicine and it is to be sold as a drug. This new regulation adversely effects all nutritional supplements, not just herbs."

Next april the European directive regarding traditional plants will regulate the use of plant based products that were freely exchanged before.

This directive requires that all preparations containing plants are submitted to the same type of procedure and have the same conformity norms as medecine.

SPECIAL FOREST PLANTATIONS

Of particular interest in many arid regions of the world are forest plantations that are established as windbreaks and shelterbelts, for sand dune stabilization, canal side and riverside plantations, and as amenity plantations.

WINDBREAKS AND SHELTERBELTS

In arid zones, the harsh conditions of climate and the shortage of water are intensified by the strong winds. Living conditions and agricultural production can often be improved by planting trees and shrubs in protective windbreaks and shelterbelts which reduce wind velocity and provide shade. Windbreaks and shelterbelts, which are considered synonymous in this manual, are barriers of trees or shrubs that are planted to reduce wind velocities and, as a result, reduce evapotranspiration and prevent wind erosion; they frequently provide direct benefits to agricultural crops, resulting in higher yields, and provide shelter to livestock, grazing lands, and farms. A main objective of windbreaks and shelterbelts is to protect the agricultural crops from physical damage by wind.

Other benefits include:

- Preventing, or at least reducing, wind erosion;
- Reducing evaporation from the soil;
- Reducing transpiration from plants;
- Moderating extreme temperatures.

Quite often, protection can be combined with production by choosing tree and shrub species that, apart from furnishing the desired sheltering effect, yield needed wood products.

DESIGN OF WINDBREAKS AND SHELTERBELTS

When considering windbreak or shelterbelt planting, three zones can be recognized: the windward zone (from which the wind blows); the leeward zone (on the side where the wind passes); and the protected zone (that in which the effect of the windbreak or shelterbelt is felt).

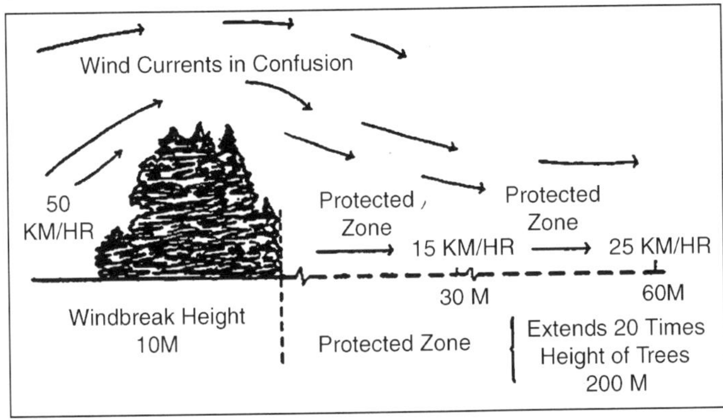

Fig. Functioning of a Windbreak

The effectiveness of the windbreak or shelterbelt is influenced by its permeability. If it is dense, like a solid wall, the airflow will pass over the top of it and cause turbulence on the leeward side due to the lower pressure on that side; this gives a comparatively limited zone of effective shelter on the leeward side compared to the zone that a moderately permeable shelter creates. Optimum permeability is 40 to 50 per cent of open space, corresponding to a density of 50 to 60 per cent in vegetation. Gaps in the barriers should be avoided. Permeability of dense shelterbelt can be improved by pruning lower branches at 0.50-0.8 m from the soil level.

It is generally accepted that a windbreak or shelterbelt protects an area over a distance up to its own height on the windward side and up to 20 times its height on the leeward side, depending on the strength of the wind. In reducing wind speeds, narrow barriers can be as effective as wide ones. Furthermore, a narrow shelterbelt has the advantage of occupying less land.

The shape of the cross-section of a windbreak or shelterbelt determines, to a great extent, the sheltering effect. To a large extent, the choice of tree or shrub species to plant, along with their planting arrangement, dictates the cross-sectional shape. In general, an inclined slope facing the wind should be avoided, as it only deflects the windflow upward. Barriers with a clear vertical side provide best wingspread reduction. When designing a windbreak or shelterbelt, the direction of the wind must be considered. A barrier should be established perpendicular to the direction of the prevailing wind for maximum effect. To protect large areas, a number of separate barriers can be created as parts of an overall system. When the prevailing winds are mainly in one direction, a series of parallel shelterbelts perpendicular to that direction should be established; a checkerboard pattern is required when the winds originate from different directions. Before establishing windbreaks or shelterbelts, it is important to make a thorough study of the local winds and to plot on a map the direction and strength of the winds.

Selection of Tree and Shrub Species

In the selection of tree or shrub species for windbreaks or shelterbelts, the following characteristics should be sought:

- Rapid growth;
- Straight stems;
- Wind firmness;
- Good crown formation;
- Deep root system, which does not spread into nearby fields;
- Resistance to drought;
- Desired phonological characteristics (leaves all year long or only part of the year).

Planting Techniques

Planting techniques for windbreaks and shelterbelts are identical to those in other tree and shrub planting programmes. However, as windbreaks and shelterbelts require a high plant survival rate, as well as uniform and rapid growth, supplementary irrigation may be required during the establishment phase. Gaps cannot be tolerated and, when plants are lost, replacement must be prompt. Although in theory, one-row barriers should suffice, experience has shown that the most effective windbreaks and shelterbelts are those consisting of several rows of trees. Quite often, initial spacing is 3 meters between the rows, with trees 2 meters apart in the rows. Where trees or shrubs have long roots that could extend into agricultural fields, vertical root pruning may be recommended; this can be done with special equipment or by digging trenches. A triangular arrangement of plants is frequently prescribed.

Management Practices

Once established, the effectiveness and longevity of a windbreak or shelterbelt depends on its maintenance. As the trees and shrubs mature, they change in shape and appearance, which necessitates some level of maintenance to ensure a continuing shelter effect. Pruning may be required to stimulate height growth, while thinning can boost diameter growth. To keep a barrier at the desired density and permeability, occasional pruning or removal of plants may be necessary. If trees or shrubs are damaged by wind or pest attacks, a control is also needed. In all of these cases, the management practices depend on the desired composition of the barrier and the species used. Since these management practices can involve the removal of woody parts, the use of tree or shrub species that make fuelwood or fodder available on a continuous basis is desirable.

A windbreak or shelterbelt has a life that is dependent on the trees or shrubs of which it is composed. Therefore, to be able to furnish permanent shelter, a renewal plan should be adopted. To renew a barrier consisting of many rows, felling the rows on the leeward side and then replanting them is often recommended. If the windbreak or shelterbelt consists of one row, a new row may be planted parallel to the the old one; when the new row has matured, the old one is removed. To renew narrow windbreaks or shelterbelts arranged into a system, new belts can be planted midway between the existing barriers which, in turn, are to be removed when the new ones become effective.

When windbreaks or shelterbelts are established on grasslands or other areas where animals are allowed to graze, special attention must be paid to the protection of the barrier; this can be done by planting thorny vegetation or by using a barbed-wire fence along the edges of the barrier.

Sand Dune Stabilization

Sand dunes result from wind erosion. They are formed in many arid lands when winds regularly blow over poorly-vegetated areas. Sand dunes that are not covered with vegetation (because of overcropping or overgrazing) move in the direction of the wind at a speed which can approach 10 meters a year, endangering agricultural crops, forest plantations, irrigation canals, and roads. To prevent this encroachment, the sand dunes must be stabilized; one method of sand dune stabilization is to establish a vegetative cover.

In general, two types of sand dunes are recognized: coastal dunes and inland dunes. Techniques of stabilizing these two types of sand dunes through the establishment of a vegetative cover.

Stabilization of Coastal Dunes

Coastal dunes originate from sand thrown up onto the shore by waves. At low tide, the sand dries and is blown away by the wind. When protective vegetation beyond the beaches is destroyed, coastal dunes move inland. To stop the advancement of coastal dunes, an artificial foredune should be constructed about 50 meters from the floodline. Normally, this initial barrier is built one year before a planting programme begins.

One method of building a foredune is by mechanical fixation of the sand by fences or palisades, 0.5 to 1 meter high. The materials used for the fences or palisades may include twigs from trees or shrubs, brushwood, grass sheaves, reeds, bushes, palm leaves, old railroad ties, used oil drums, and earth. When the prevailing wind has a prevailing direction, parallel lines of palisading are sufficient; however, a checkerboard system is advisable where fluctuating winds are common. Sand piles up behind the palisade and, when the artificial dune that is formed reaches a height of 0.5 to 0.75 meter, a second palisade is built on top of it. Sometimes, the original barriers can be raised, when necessary, instead of building a new palisade.

Once the foredune is established, it is possible to stabilize the sand behind it by seeding or planting a vegetative species that provides good ground cover and is able to withstand (at least partially) covering by sand.

Sand dune fixation also can be done by mechanical mulching; that is, the spreading of solid material on the surface of the sand. Chemical fixation can also be employed. Chemical fixation consists of stabilizing the sand surface by covering it with a continuous crust of sprayed chemical substances, such as petroleum derivatives or latex mixtures. Vegetative establishment is usually a follow-up or a concurrent operation. Chemical fixation is advisable when the cost of labour is high and the chemicals are readily available.

Sand dune stabilization with plant species is more permanent than mechanical mulching and chemical fixation techniques which are, in most cases, only temporary measures.

Stabilization of Inland Dunes

Inland dunes originate from sand produced by the weathering of rocks, mainly sandstone. The fine fraction can be blown far away, while the heavier fraction is blown short distances and forms dunes. Such dunes can pose serious stabilization problems, especially when the dunes are large and active. One way to combat this problem is by creating an artificial dune at the windward end of the dune. The method followed is similar to the one used to create the foredune in the stabilization of coastal sand dunes. The stabilization of inland dunes also follows the same general lines.

When a specific area of value (for example, an oasis) is threatened, protective work is initiated as close as possible to the area of concern, with the work gradually progressing towards the sand source area.

Planting Techniques

Planting of vegetation is the best and most permanent method of coastal and inland sand dune stabilization; both direct and indirect benefits can be realised, including:

- Protection (of roads, canals, agricultural lands, and industrial areas);
- Wood production (fuel, lumber, etc.);

- Protection of watershed areas and water supplies;
- Livestock grazing benefits (including fodder);
- Wildlife benefits, recreation, and other amenities;
- Public works to combat unemployment.

The choice of vegetative species for planting should be based on studies of the natural vegetation in the area and on the environmental conditions. As planting of vegetation on sand dunes frequently consists of afforestation practices, it is recommended that species trials be included in the planting programmes to permit an evaluation of tree and shrub species for long-term use.

In practice, it is often necessary to plant relatively large containerized plants close together (1 × 1 meter) on the windward side, but they can be planted further apart (2 × 2 meters) on the sheltered side. Irrigation for initial establishment may be required to help the plants survive until they have sufficiently deep root systems. If water is not available in adequate quantities for irrigation to take place on a long term basis, it is advisable to irrigate (at least) during the first two or three months after planting, at weekly intervals.

Concerning maintenance, hand weeding is preferred to avoid problems of machinery traction in the sand. As a rule, all livestock movement and other traffic should be eliminated on the sand dunes; when necessary, delimited and protected passages for livestock can be established.

CANAL–SIDE PLANTATION

In may arid countries, wherever rivers are available, efforts have been made to utilize the water for irrigation purposes through the construction of dams or using lift irrigation for the agricultural needs. Several thousands of kilometers of irrigation canals have been laid. The banks of such canals are available for planting purposes and constitute a considerable area for production of timber and firewood for the rural population. Full advantage is being taken of this in many countries like China, Egypt, India and Pakistan. A few rows of trees, varying from 4 to 6, are generally planted on each bank of the canal with an espacement depending on the characteristics of the species and the type of produce desired.

When designing a canal plantation, the requirement may be the same as for the design of irrigated plantations with respect to climatic and soil conditions and to supply and quality of water. However, it should be remembered that the only water supply available to the trees is seepage from the canal into the root zone. In some places, it is cheaper to grow trees and thus utilize the seepage water rather than prevent seepage by canal linings of concrete, asphalt or other material.

Choice of species for canal side plantations should take into account both the particular character of the plantation and its purpose. The roots of the trees should strengthen the banks of the canal and the trees should keep the canal and its banks well shaded in order to suppress weed growth and reduce evaporation. Species that tend to increase water seepage through the sides and bottom of the canal should be avoided. Where canals have an intermittent flow, such as flood discharge canals, only trees able to adjust to varying water levels in the soil can be used.

Species that reproduce by suckers such as Robinia pseudo acacia should not be planted along canals. Plantation techniques should favour deep planting and roots should be planted in the moist layer.

RIVER-BANK PLANTATIONS

There are many areas where river lengths are considerable. The ground on either side of the river is partly within the reach of the high level of water during the period the rivers are in flood. Beyond this level—and on the fringes of the agricultural land, strip plantation can be established to produce wood, fuelwood and fodder. Generally, the width of such strips is limited but does constitute a useful and productive linear plantation. Underground water is available at different levels. The species to be planted should be matched with this water level variation.

Spacing within and between the rows depends on the characteristics of the species and the rotation planned for the crop. In the more arid areas, trees with xerophytic habit constitute the outermost rows while those close to the river bank are the ones with higher water requirement. In such locations, phreatophyte species such as Populus spp., Acacia nilotica, Dalbergio sisso, Prosopis spp. can be planted.

AMENITY PLANTATIONS

This type of plantation includes trees planted in gardens and parks, street planting, green belts around villages and cities, trees planted along roadsides to reduce noise and beautify the homestead or landscape. In arid zones "beautifying the landscape" usually means changing the countryside from its normal brown colour to green, or "greening" of the landscape.

Tree Planting in Gardens

Tree planting in gardens is usually regarded as a beautification of the home environment but it also has a profound effect on man's psychological attitude towards life. A house located in a barren landscape without trees and shrubs lacks appeal and there is certainly a different psychological attitude towards it, compared with a home which has been beautified and protected by wisely selected trees and shrubs. Tree planting in gardens also enhances self-esteem. The gardener identifies with his garden and builds a personal relationship with it.

The garden becomes an extension of himself, a visible representation of his individuality. When it blooms, he has evidence of his success. He also becomes aware that a number of people he does not know pass by each day and enjoy his garden. He has given them an anonymous gift. All of this enhances his self image, helping to create self-esteem. The gardener, feeling better about himself, feels better about where he lives.

In selecting species for tree planting in gardens, the following should be considered:

- Trees have several functions to play in plantations around houses. The first point to be considered is therefore to decide on the purpose for which planting is to be done, that is: shelter, shade, ornament, hedges, scent and odours, as source of fruits or nesting sites for birds. Some plants will fill only one or two of these objectives while others may fill more.
- Trees selected should be suitable to local climate and soil conditions.
- Adaptability to pruning. The degree to which trees and shrubs will tolerate pruning and pollarding is an important consideration for planting around the homestead.
- Evergreen or deciduous habit: most arid-zone species are evergreen but many deciduous trees are also used for planting in gardens. In the Mediterranean arid region, deciduous trees have certain advantages and disadvantages when compared with evergreens. The most important advantage is their ability to provide shelter in summer but allow sunlight to penetrate in winter. The most important disadvantage is that the leaves need to be gathered during and after leaf fall.

Planting in Parks

The first objective for planting in parks in arid zones is for shelter from the sun and dust. Where trees are not growing naturally, shelter may be provided by planting trees in favourable sites. A second objective is to enhance the beauty of the park by planting trees of different colour, shape and size. Planting formal layout patterns of trees in long straight rows does not usually fit into the landscape pattern, and planting in small clumps or as individual dispersed trees is usually more suitable. Choice of species should not be dictated by the value of trees for wood, nor planting expenses weighed against the value of the wood produced. Costs for planting in parks for recreation should take into account aesthetic benefits and the intangible gains in the health and well-being of the people.

Street Planting

Street plantings are often the responsibility of local municipalities and are made to beautify the cities, provide shade and control outdoor noise and traffic pollution. In recent years, many of the towns and cities in the world have learned through experience that paving roadways, streets, and sidewalks does not complete the job. In most communities where there has been an increase in population and greater congestion of motorized traffic, it has become important that the city municipality assumes its responsibility of providing the amenities which add much pleasure to life in the city.

For use in street planting, trees should:

- Be easy to establish, preferably with the ability to be transplanted as advanced nursery stock, and grow relatively quickly to the stage that they provide some amenity value;
- Be healthy in the environment, relatively long-lived and not subject to wind-throw or the breakage of large limbs;
- Be as maintenance-free as possible. Trees requiring permanent pruning and removal of fallen leaves will have a high maintenance cost.
- The form and height of the species must be suitable for the width of the street in which they are to be planted;
- Whether to use a single variety or mix on the same street is a matter of taste. A variety might be used depending on the geographic section of the country and the width of the street.

Greenbelt Planting

Several cities in arid zones have established in their municipal area green belts with a number of purposes:

- To enhance the beauty of the site;
- To provide a recreation area for the urban dwellers;
- To reduce the harmful effect of dry winds and dust storm and control sand encroachment.

There is a wide array of trees and shrubs for greenbelt planting. Any programme of this sort, however, should be well designed in advance, planned for a number of years and carefully implemented.

Roadside Plantations

Roadside plantations have several objectives:

- Trees increase the comfort of travellers by providing shade and attractive surroundings;
- Trees may protect the road itself against moving dunes or act as a windbreak for adjacent fields.
- Trees may become an important factor by alleviating timber and fuelwood shortage. In fact, roadside trees are frequently considered a part of the national forest planting programme. Such trees may produce edible fruit, yield pods for feeding animals, furnish food and shelter for birds or, when in bloom, be valuable in beekeeping.

Species should be carefully chosen. Among the important factors which should be considered in planning the use of trees along highways:

- Selection of the species for hardiness, longevity, freedom from windthrow and breakage, attractive appearance and minimal maintenance. In the arid zone there are a large number of native small trees and large shrubs which can be used, as well as some exotic species. Where the environment is suitable, consideration should be given to small patches of deciduous species with colourful foliage.

- Suitability of the species to the climate, topography and soil.
- Location of the trees in relation to road formation. Firstly consideration should be given to the existing road formation so that trees ar not planted close to the inside of curves or near road junctions where they could obscure vision and so create a driving hazard. Secondly, consideration should be given to the possibility of the future widening of roads, including the development of double traffic lanes.

Rail-side Plantations

A trend towards rail-side planting for the provision of greenery, protection from dust and winds and creation of additional tree resources has developed in recent years in many countries. This trend is likely to spread to other countries due to the favourable results already achieved in certain countries. Three to six rows of trees on either side of the track are considered useful. The planting techniques are similar to those for roadside planting. Species vary and depend on the prevailing climatic conditions, mainly temperature, soil and rainfall.

Under very arid conditions, the choice of species is rather limited. Where water is available, several species can be selected. Within an area each row can be given over to one or more species. Mixtures are thus created and are considered better as they yield different produce to meet local needs.

9

Challenges of Community Forestry in India

Forests have been providing substantial support to the rural economy in Asia. Historically, the local communities managed their forests for the supply of fuel, fodder, fibre, timber, food and herbal medicines while maintaining an ecological balance. However, with the biotic pressure from growing human and livestock populations, lack of technical skills, poor investments and change in the ownership, forests in India have depleted rapidly over the past five decades.

This led to the involvement of rural communities in forestry development during the early 1970s. The primary objective of community forestry was to generate employment, protect the environment while ensuring basic needs of fodder, fuel and timber. However, the schemes could not fulfill the objectives due to lack of people's participation. Hence the community forestry during the next decade should focus on enhancing the productivity of natural resources, while empowering the local communities.

Development of tree based farming systems in association with watershed development is expected to be a popular community forestry programme in the near future in India. There will be a preference for tree species producing fruits, timber and non-wood forest products because of sustainable income. As wood will continue to be the major source of fuel in rural areas, efforts will be made to produce woodfuel by introducing fast growing tree species in mixed plantations. Development of community wastelands, rehabilitation of mined areas, protection of natural forests, establishment of industrial greenbelts in urban area and environmental awareness are the other important components which will gain significance. Development of local organisations for input supply and processing of the produce, marketing and information sharing, based on the research studies require further emphasis for effective management of the programme. Research back up is needed to improve the productivity through selection of improved germplasm, enhancement in growth and yields, value addition through processing and marketing.

BACKGROUND

About 43% of the energy consumed in the Third World is being met from wood. In India, biomass constituted 85% of the rural energy and with a per capita consumption of 1.0 ton/year, about 50% of it was collected from forests. Apart from supplying wood, forests have also helped in conserving biodiversity, agricultural productivity and a safe environment. Historically, forests provided substantial support to the rural economy. Over 60% of the Indian population was directly dependent on forests for fuel, fodder, fibre, timber and a wide range of food and medicinal herbs. The forests provided year round employment to 20 million people through collection of Non-Wood Forest Products (NWFP). For over 50-60 million people representing 250 tribal communities, forests formed a part of their culture and a natural way of life.

In India, out of the total land area of 329 million ha, 143 million ha were under agriculture and 77 million ha were classified as forests. The country had a distinct and varied vegetation from the tropical wet evergreen forests of Andaman and Nicobar islands to the dry Alpine forests of Himalayas. However, with only 2% of the world's forests, there was severe pressure in serving 15% of the world's population. The woodfuel consumption during 1970-80 was around 175 million tons/annum and most of it came from forests and village common lands. Apart from fuel, forests provided 45-60 million m^3 timber and 45-60 million tons of forage annually.

It was the traditional wisdom and joint responsibility of the communities that enforced necessary rules and regulations on the local people for sustainable management of their forests. However, with increasing growth of human and livestock populations and a shift in the ownership of natural forests from the princely states to the Federal Government, the control and moral pressure on the local population was relaxed, leading to indiscriminate abuse of forests since the middle of this century. Unable to meet their basic needs from agriculture, many unemployed and poor families turned to forests not only for fodder and fuel, but also to generate cash income through sale of wood and other forest products. Vested interests also took this opportunity to exploit forests for commercial purposes.

Apart from biotic pressure, the following factors contributed to the denudation of forest resources (GOI, 1984).

- Increase in human and livestock populations;
- Poor management of forest soils;
- Inadequate scientific and technical inputs;
- Inadequate skills and training of the staff to play their expected new roles;
- Poor investment on forest development;
- Damage caused by mining, irrigation projects, industries, roads and shifting cultivation.

The abuse of forests continued unchecked till mid 1970s, when the ill-effects of deforestation were prominent in the form of fodder and woodfuel scarcity, soil erosion, flash floods, water scarcity, loss of precious flora and fauna and climate change. By 1981, the area under wastelands in India was estimated at 93.70 million ha, excluding the degraded agricultural lands measuring over 85 million ha. Most of the natural forests around the villages had turned into wastelands reducing the area under forests to only 40 million ha. Over 13 million ha of community pastures were devoid of vegetation due to over-grazing by 450 million heads of livestock. Due to the non-availability of wood on village common lands, rural women were compelled to spend 15-35 hours every week in walking long distances for collecting fuelwood from interior forests. As a result of deforestation, the damage from floods affected 58 million ha of agricultural land and over 60 million people during the 1980s. The extent of damage had increased by four folds over the earlier two decades.

The rural poor, particularly the women, who are primarily responsible for fetching water, fodder and fuel were faced with severe drudgery. The community felt that while it was the responsibility of the Forest Department to manage the forests, it was their inherent right to collect fodder, fuel and other products from the forests, without any obligation. Planting of trees for fodder and fuel was never considered by the farmers as a necessity. Meanwhile, faced with shortage of biomass, many village communities resisted the extraction of wood from the forests for commercial purposes. This led to the involvement of rural communities in forestry development programmes in India.

COMMUNITY FORESTRY

In 1976, the National Commission on Agriculture in India introduced the concept of Social Forestry to encourage those who were dependent on fuelwood, fodder and other forest products, to meet their own needs through various activities, in order to reduce the burden on the Forests. This concept was further refined by FAO in 1978, by defining community forestry as the programme which intimately involved local people in afforestation, irrespective of the pattern of land ownership. While the traditional forestry covered the protection and production roles, social forestry was intended to play the social role.

Thus community forestry programmes included a wide range of activities such as growing trees on farm bunds and roadsides, developing woodlots on common properties and collection, processing and management of forest products by involving local communities. Realising the problems of poverty and unemployment closely linked with denudation of natural resources, scarcity of fuel, fodder and timber, the Government of India in their Sixth Five Year Plan (1980-85) aimed at generating rural employment through social forestry and thereby reducing poverty from 48% to 30%, which was further intended to be brought down to 10% during the Seventh Five Year Plan (1985-90). It was also intended to produce adequate fodder and fuel to meet the needs of the poor, particularly in forest deficit areas. The major focus was on the extension of forestry programme to non-forest lands with the active involvement of local communities. This was essential because only about 40% of the 5 million villages had access to forests, while the rest had to look for alternate sources of woodfuel. It was also an opportunity to develop 175 million ha of degraded lands. Subsequently, in the 1980s, the scope of social forestry was enlarged to cover the protection of reserved forests and establish tree species on vacant sites in urban areas. However, the programme did not include large-scale industrial forestry.

In 1985, the National Wastelands Development Board was established by the Government of India to promote the production of fodder, fuel and minor timber on wastelands, by involving local communities and voluntary agencies. Important social forestry models formulated to achieve these goals were - rehabilitation on degraded lands, strip plantations, village woodlots, farm forestry, agroforestry, homestead plantations and decentralised nurseries. Newly established State Social Forestry Departments were confident of implementing this programme, in view of their strong technical expertise and adequate resources. However, in the absence of social mobilisation, involvement of the local people was confined only to wage earning on community woodlots, while the implementation of these schemes was carried out by the Forest Officials.

Community Forestry Schemes in 1980's

By 1990, the drawbacks of various social forestry schemes launched in different parts of the country became apparent. The Social Forestry Departments had great difficulty in mobilising local participation. There was a wide gap between the assumed problems and actual realities in the field. As the work plans were not based on specific ground data and ecological conditions, the proposed models of social forestry were often found to be inappropriate. However, modifications in the project designs and implementation plans could not be carried out because of rigid rules and regulations. These schemes could not register the anticipated success because of the following reasons :

Community Woodlots: Development of fuelwood plantations on community wastelands was a major programme to generate employment for the local poor and the landless, while augmenting the fuel and forage shortfall faced by the community. However, this scheme could cover only 9% of the villages either due to non-availability of wastelands or unwillingness of the Village Panchayats and public institutions to spare their wastelands for community plantations. Many of these projects had failed to progress in terms of plant survival, growth and income generation. In the absence of effective local organisations, the programme heavily depended on the Village Panchayats, who had bare minimum staff to manage these plantations. As the area available for plantations was very small, the supply of fuel and fodder was also inadequate to meet the local demands. It was also realised that the problems of fuelwood and fodder shortages could not be solved simply by promoting the cultivation of these species, as these commodities were of least value as compared to timber wood. Therefore, despite the shortage of fuelwood, the farmers preferred to grow timber and fruit species.

In some successful community plantations, inspite of the offer to collect grass free of cost, the response from farmers was very poor. The villagers felt that feeding such fodder to poor quality animals would not improve the productivity except in a few locations where they had maintained livestock of improved breeds. As the community plantation activities could not motivate the local people, most of these plantations were destroyed within a short period after completion of the project.

Strip Plantations: Saplings of tall growing tree species were established in multiple lines along the railway tracks, canal bunds and roadsides and the landless families were involved in protecting them. This was fairly successful but the incentives offered by the project implementing agency were not adequate for the beneficiaries to devote sufficient time for protection. Thus some of these trees were stolen. In the absence of clear directives from the government regarding the sharing of benefits, the participants were not allowed to cut the trees.

Farm Forestry: It was the most successful scheme wherein farmers established plantations of eucalyptus, casuarina, poplar, teak, etc., on their agricultural lands using necessary inputs like irrigation and fertilisers with the hope of finding a suitable market and attractive price for the produce. Many of these farmers located near the wood based industries could sell wood at remunerative prices. However in the absence of any marketing infrastructure it was difficult for them to find attractive buyers in remote areas. Obsolete Tree Acts imposed by the government created further obstacles in felling, transporting and sawing of timber by the growers. This forced them to either sell it to the middlemen or delay the harvest.

The community forestry programme also included a component of planting fruit species on marginal lands owned by poor families belonging to scheduled castes and scheduled tribes. However, the project could not get adequate response at the initial stage as most of the illiterate target families feared that the government would acquire their lands, after the establishment of trees. Subsequently, even the limited number of participants in the programme could not maintain the plants due to inadequate resources to provide irrigation and protection.

It was also observed that while promoting farm forestry, farmers were not informed about the profitability of various tree species which could be grown under local agro-climatic conditions. As a result, the farmers preferred only a few species such as eucalyptus, which they thought to be most profitable, while many other fruit and non-timber forest product species which could have provided higher income were neglected.

Decentralised nurseries: The primary objective was to build up the local capacity for assured supply of planting materials of popular tree species. As the focus was on fodder and fuelwood plantations, the participants were advised to raise those saplings which were of no interest to them. In the absence of well planned marketing arrangements, the nurserymen were totally dependent on the Social Forestry Department for selling their seedlings. As a result, farmers had to discontinue their activities immediately after completion of the project.

Energy Conservation: Several energy conservation measures like promotion of improved wood stoves and biogas plants did not achieve the expected results due to poor publicity and awareness. The poor response was also because of unsuitable designs of smokeless wood stoves which failed to save wood.

Lessons Learnt

No doubt the community forestry programmes in 1980s could not prove to be people-oriented, but they did generate a lot of awareness among the local people. Many success stories were recorded at micro-level which were good for replication and motivation of the rural people on a larger scale. Several lessons were learnt by the Project Implementing Agencies.

Although biomass was the major source of fuel for cooking, the supply was not critical. Contrary to the earlier belief, it has now been understood that only 30 - 40% of biofuel was in the form of wood and only 10 - 15% of it was from natural forests. The rest of the biofuel was from crop residues, animal dung, shrubs and bushes grown on non-forest lands. However, as the demand for woodfuel will increase at 1.5 - 2.0% per annum, the estimated consumption of woodfuel in 2010 will be 226 million tons against the supply of 256 million tons. Apart from the wood for fuel, the annual demand for wood from industries, housing and agricultural sectors is 125 million tons, leading to further pressure on forest resources.

Compared to other sources of energy, woodfuel has several advantages. Firstly, wood can be easily regenerated through the establishment of energy plantations on marginal and wastelands. With its tropical climate, fairly good quality soil and well distributed rainfall, India presents an ideal condition to produce 10-15 tons of biomass per ha/year. Woodfuel production and distribution can also provide employment for rural people at various levels. The woodfuel business can provide employment for about 10% of the rural households and contribute to about 40% of their total earning. Inspite of huge involvement of labour, energy from wood is cheaper than fossil fuels.

Woodfuel is also environment friendly, when the rate of biomass harvest is equivalent to the rate of regrowth. In such a situation, CO_2 released from wood while burning will be absorbed by the greenery without any burden on the environment.

The use of woodfuel in the Asia Pacific Region in 1994 has resulted in saving 278 million tons of CO_2, which otherwise would have been released into the global atmosphere. With continued use of wood as a hypothetical replacement for LPG, the annual CO_2 saved will be 349 million tons in 2010. The estimated cost avoided by recapturing the CO_2 is estimated at US $ 14 billion in 1994 and US $ 17.5 billion in 2010. Thus, biomass production deserves high priority in community forestry.

Realising this need, emphasis was laid on Joint Forest Management which involved people's organisations to launch tree plantations both on forest and non-forest lands. This programme emphasized on the formation of Village Forest Committees and capacity building of the committee members to take suitable decisions on afforestation and its protection without any interference from the donor agencies and Forest Department.

Inspite of its good intentions, the success of the Joint Forest Management Programmes was also dependent on other factors such as productivity of the plantations based on soil fertility and composition of the tree species, demand for the produce, political support and transparency of the project implementing agency. In some states where the government had issued clear directives on the sharing of benefits between the government and local people, the latter had shown greater enthusiasm, resulting in a higher degree of success.

The success of the community forestry programme was also influenced by political commitment, correct assessment of local needs, appropriate technology, system of providing incentives, suitable local organisations and support services for finance, management, research and extension. Inspite of major setbacks, the Government of India has emphasized that those community forestry programmes which aim at conserving the natural resources should address the problem of poverty.

It was realised that the community forestry could not solve all the problems of the rural communities, although it had good potential to support the rural economy. There was a need to involve the local communities for planning the programme till they developed suitable interventions to fulfill the goals.

The key task for successful implementation of the programme in the future was to understand the socio-economic aspects of the local population and design a suitable technical programme to ensure sustainable development. The programme should have a strong component of extension for motivating the participants. The extension programme should highlight the benefits of various schemes through effective media, such as radio, TV, documentary films and newspapers. However, as most of the farmers were illiterate, involvement of local leaders, members of the village *Panchayats*, progressive farmers, school teachers and voluntary agencies was most effective in ensuring better people's participation and adoption of new technologies.

PROGRAMME PRIORITIES FOR THE FUTURE

The community forestry programmes during the next two decades should focus on the following aspects.

- Enhancing the productivity of natural resources, while augmenting the basic needs of the community;
- Empowering the local communities to initiate the process of planning and programme implementation at the microlevel to promote afforestation for income generation and ecological conservation.

The programme during the next two decades may be divided into two phases. During the first phase of 10 years, emphasis should be given to enhance the productivity, while conserving biodiversity and the environment. The second phase should consolidate the programme by promoting micro and macro-level enterprises, involving the local people and the business interests to sustain the development and optimise the opportunities for employment and income generation.

AGROFORESTRY FOR IMPROVING PRODUCTIVITY OF PRIVATE LANDS

As over 70% of the agricultural lands in India are dependent on rainfall, the crop yields are comparatively low. In such a situation, promotion of tree based farming system has good potential to improve the Land Equivalent Ratio (LER) as well as crop yields. The system of combining horticultural crops with fodder, fuel, NWFP and agricultural crops can ensure the supply of essential commodities while generating cash surplus. Among these tree species, the income from fruit and NWFP plantation are generally higher than timber and industrial wood plantation, because of substantial annual income without cutting the trees.

Development of tree based farming systems on the watershed basis has the potential to improve water supply, food production, employment generation apart from supplying fuel and forage. Water conservation can be facilitated through contour bunding, gully plugging, formation of farm ponds and percolation tanks. This programme can help in recharging the ground water table to provide water for human beings, livestock and crops. The selection of a suitable farming system will be based on the soil fertility and moisture availability. As the Government of India has already launched a massive programme of watershed development, linkage of agroforestry as an integral part of this programme will not only provide an opportunity to enhance biomass production, but also improve the profitability.

While fruit orchards are ideal for deep soils having some water resources to protect the plants in the initial stages, farm forestry and agri-silviculture are suitable for marginal lands. With introduction of fruit species, farmers can be motivated to intensify their farming operations and improve the soil productivity. Through soil and water conservation, green manuring vermi-composting, agroforestry and integrated crop protection, the yields of intercrops can be substantially high, atleast for the next 10-15 years, till the intercrops are affected by the shade, although the LER will remain significantly high to maintain the orchards. Such tree based farming systems have been highly effective in preventing migration of rural families and supporting livestock husbandry to enhance their income by 35-40%.

Wood Production and Conservation

Commercial Plantations: Wood production should continue to be the priority under community forestry. However, fuelwood being a low value commodity, farmers prefer to grow superior quality wood suitable for timber and industrial use. Therefore the priority should be to promote cultivation of timber and industrial wood on degraded agricultural lands and community wastelands, which are not suitable for cultivation of fruit and NWFP species. The wood based industrial and housing sectors which are dependent on natural forests can associate with farmers to procure their raw materials. This programme can be managed with least financial support from the government, as commercial banks and industries are prepared to provide necessary finance.

Mixed Plantations: While the industries utilise the wood of a certain minimum diameter, almost 40-50% biomass in the form of branches and twigs can be available for fodder, manure and fuel, almost free of cost. Apart from producing fuelwood as a byproduct in commercial plantations, fast growing fuelwood species of short gestation can be introduced on field bunds and as mixed crops with commercial wood species. This will augment the fuelwood supply to a great extent.

Wood Saving Devices: Efforts are also needed to introduce various energy conservation measures. These may include the development of fuel efficient, low cost wood stoves, promotion of improved cooking vessels and introduction of efficient models of biogas to substitute woodfuel for cooking. Biomass being a renewable form of energy, it can be an excellent source for generating electricity in wood surplus countries. However, it is necessary to develop efficient gasifiers to take advantage of this technology in the future.

DEVELOPMENT OF COMMUNITY WASTELANDS AND DENUDED FORESTS

Conservation of biodiverstity and protection of forests are the other important components of community forestry. However there has been a lukewarm response to these activities in the past, mainly because of lack of awareness, inadequate benefits and lack of trust between the local communities and government agencies who own these lands. Hence the new strategy should find a suitable solution for these problems.

Development of forest lands: Rehabilitation of denuded forest lands and protection of natural forests are essential to conserve biodiversity. The forests influence the climate and soil moisture conditions and improve agricultural production.

This aspect needs wider publicity in rural areas and the local people should be motivated to take active part in Joint Forest Management. The Village Forest Committees willing to protect the natural forests, should be encouraged to share the benefits in the form of fodder, fuel, timber and NWFP. These committees can also take up reforestation of degraded forest lands, aiming to conserve biodiversity and generate income. While facilitating natural regeneration, the local communities can also introduce fuel, timber and NWFP species of their choice to enhance the productivity on denuded forest lands.

Development of community lands: Lack of people's initiative was a serious problem in developing community wastelands, mainly due to lack of direct benefits. Earlier efforts to establish woodlots could not generate substantial income. Establishment of NWFP species such as tamarind (*Tamarindus indica*), Indian gooseberry (*Emblica officinalis*), neem (*Azadirachta indica*), Pongamia (*Derris indica*), *Madhuca spp*. and a wide range of medicinal species can generate substantial income, regularly for over 50-100 years. With the world's herbal medicine business likely to cross US $ 3 trillion during the next decade, it will be an excellent opportunity for the Asian farmers to enhance their income through cultivation of these species on community lands as well as on private wastelands.

Development of Mined Areas: Rehabilitation of mined areas is another cause of serious concern. Although the industries are obliged to establish a green cover after mining the area, the outcome has been very poor. Such sites are major causes of soil erosion and flooding of rivers. Lack of commitment, inadequate expertise in selection of plant species and their management have been the main reasons for failure. These activities can be entrusted to voluntary organisations and local communities, with research support to raise healthy plantations. Preference should be given to NWFP and timber species which have long gestation period, sufficient to enrich the mined lands.

CO_2 Sinks: There are opportunities for environmental protection by establishing industrial greenbelts, recycling treated effluent and developing carbon dioxide sinks, particularly around urban areas. The polluting industries can be compelled to establish greenery through the involvement of voluntary organisations and local people.

Environmental Awareness and Eco-tourism: Lack of awareness among the common public is a cause for environmental pollution in the developing countries. Hence, continuous efforts are needed to motivate the common public to plant more trees and protect the environment. School children and literate urban population are fairly receptive to participate in environmental protection. This can be undertaken through school based plant nurseries, development of arboretum, botanical gardens and greenbelts around the schools and industries.

While developing industrial greenbelts and afforestation on river banks, special efforts can be made to promote eco-tourism. Creating waterbodies through plugging of gullies, developing, nature trails, shady grounds for resting, selecting suitable plant species to attract wildlife and providing adequate eco-tourism. Environmental exhibitions at these sites can help in motivating the public to take active part in community forestry and environmental protection.

ORGANISATIONAL DEVELOPMENT

Development of local organisations is the key to ensure people's participation and to sustain the benefits of investments and efforts made in community forestry.

Tree Growers Organisations

The initial need is to provide leadership to organise local people in afforestation. This can be carried out through formation of local Selp-Help Groups, village level farmers' associations and forest groups. This activity should receive priority during the next decade.

Capacity Building

Training of local farmers, supply of critical inputs, credit and technical advisory services to improve the profitability and developing market networks are the critical inputs which have to be introduced in the field to support tree growers.

Networking of Tree Growers and User Groups

There is an urgent need to set up networks of tree growers, forestry users and concerned scientists in these fields. Such networks should identify the opportunities for tree growers. They can also promote processing and marketing of forestry products and protect the interests of tree growers.

Forestry Information Service

Setting up of information centres to identify the priority of the industries and farmers and provide necessary information to enhance the economic viability of community forestry would help to sustain the interest of tree growers. Forestry Extension Units should closely associate with Information Service Centres to identify appropriate technologies for transferring to the field.

Forestry Research

The networks should identify the problems affecting the production and marketing of the produce and encourage the Forestry Research Institutions to take up relevant scientific studies to solve the problems. The researchers should closely associate with the network and Information Centres to disseminate new technologies to solve field problems and enhance the production.

Forestry research should focus on the following topics:

- Conservation of biodiversity;
- Improving the productivity of wastelands, tree plantations and natural forests;
- Exploring the utility of multipurpose tree species and NWFP;
- Developing of alternatives for conserving wood;
- Selecting various organisational systems to manage community forestry.

The outcome of these research topics can further strengthen the community forestry programme in the country.

10

Soil Erosion and Conservation

Environmental problems are by no means new to American agriculture. The Dustbowl episode of the 1930s was but one of several historical instances in which concern has been raised about the environmental performance of U.S. farmers. Nonetheless, current controversies about the management of agricultural resources in the United States date largely from the rise of the environmental movement in the 1960s and have included issues relating to soil erosion, the use of agricultural pesticides, energy and water consumption, and the like. The soil conservation implications of American agriculture have remained among the most longstanding agricultural-environmental issues of concern to rural sociologists and are given principal emphasis in the section that follows.

The majority of U.S. farms are family proprietorships in which there is household or family ownership and management of the land and water resources. Accordingly, the logical point of departure in understanding the soil erosion problem in agriculture has been to conduct micro-level research on the individual, farm firm, and ecological factors associated with soil erosion rates or the utilization of soil conservation technologies. This has been a productive line of research, but it has also become a controversial one, with the major controversies centering around theoretical approach, the relationships between micro and macro levels of analysis, and the public policy inferences that should be drawn from social science research.

CONCEPTUAL ISSUES IN RESEARCH ON SOIL EROSION AND CONSERVATION

One of the most crucial aspects of soil erosion research is the very conceptualization of why soil erosion is a socioeconomic problem. Traditionally this has not been an issue, since sustained high levels of soil loss were presumed to irreversibly degrade the productivity of agricultural resources. Accordingly, if the soil erosion problem is conceptualized as medium to long-term land productivity decline detrimental to the interests of farmers, it would follow that conservation should be achieved through voluntary farmer compliance (rather than through mandatory regulation) based on the logic of the long-term interests of farmers.

Historically, federal and state government policy thus has emphasized education of farmers and modest levels of financial inducements (such as Agricultural Stabilization and Conservation Service cost-sharing programmes) to lever farmer decision making. But it has been increasingly recognized that the proportion of prime U.S. farmland that stands to be irrevocably degraded by soil erosion is relatively small and that the "off-site" costs of soil erosion — the impacts of soil erosion and run-off on water and land resources of a farmer's parcel — are very significant and may very well be in excess of the on-site costs.

These alternative conceptualizations of the soil erosion problem — productivity loss versus destruction of off-site water and land resources — have very profound implications for appropriate types of social science research and for public policy. To the degree that soil erosion primarily results in land productivity losses, research should focus on individual- and farm-level factors that influence resource management, and public policy should incorporate these research findings in order to develop educational and incentive programmes to achieve soil conservation through farmer self-interest. To the degree that soil erosion is primarily a problem because of off-site costs, farmers cannot be expected to conserve because the long-term productivity benefits of conservation are small, and research at a more macro level (*e.g.*, to determine the degree to which farmers' incentive structures and resource management behaviors are congruent with the public's interests in clean, navigable waters) would be most appropriate.

Debate over the appropriate kind of analysis in soil conservation research has been closely related to, and in many respects was stimulated by, debate over the adoption-diffusion approach. Pampel and van Es, reported that the correlates of adoption of conservation technologies tended to be different from those of commercial nonconservation technologies. They also suggested that effective conservation practices will tend not to be profitable for farmers, so that to achieve the public interest in soil conservation voluntary compliance among farmers may be insufficient. The issues raised by Pampel and van Es have continued to pervade the erosion and conservation literature. Many researchers have conducted their research by seeking to revise the microsociological diffusion of innovations approach, while others have argued that the constraints to soil conservation behavior tend to be more macro in nature and that the diffusion of innovations approach will be inadequate.

Prior to 1970 soil and water conservation technologies and practices were generally understood to be a diverse collection of cropping patterns (crop rotations, contour planting, strip cropping, cover cropping, sod waterways, filter strips) and physical and biomass structures (terraces, sediment retention basins, diversion channels, animal waste structures, and hedge rows). Since that time, however, soil and water conservation has, for the majority of researchers and policymakers, come to be largely conterminous with the adoption and use of no-till and related conservation and reduced tillage equipment, which enables farmers to leave a mulch layer of residue on the soil to reduce runoff and sediment losses.

Two observations can be made with regard to traditional and reduced-tillage equipment approaches to soil and water conservation in agriculture. First, these two approaches, while by no means incompatible, are nonetheless quite different and will tend to apply most appropriately to different types of farm operations. Traditional cropping practices and structures are often most attractive to smaller, more diversified farmers, while reduced tillage tends to be most attractive to larger, more highly mechanized producers of row crops.

Second, reduced tillage, although acknowledged to be effective in reducing erosion in row-crop monocultures, has become somewhat controversial. One controversial aspect of the reduced-tillage approach has been that it requires increased use of herbicides, which may lead to contamination of water and thereby negate some of the runoff reduction benefits of reduced tillage. Many studies have indicated, moreover, that reduced tillage practices have been adopted by farmers largely because of labour savings rather than because of the ability of the technology to reduce soil erosion.

Indeed, the farmers who have been most likely to adopt no-till and other reduced tillage technologies have been found to be large operators with highly specialized operations involving practices such as continuous cropping and row-crop monocultures. Thus farmer-adopters of such conservation technologies ironically represent the major trends in U.S. farm structure that have been demonstrated to lead to environmental degradation in agriculture. However, farmer perceptions of a threat to soil productivity may lead to adoption of soil-conserving innovations, whereas stewardship related motivations may have little effect on motivations to prevent other kinds of consequences such as off-site pollution.

SOIL EROSION AND ITS CONSERVATION

Soil is one of the most important natural resources of man. Soils are essential for man for growing crops, fodder and limber. Once the fertile portion of the earth's surface is lost, it is very difficult to replace it. In India, the destruction of the top-soil has already reached an alarming proportion. Land degradation problems have resulted in increasing depletion of the productivity of the basic land stock through nutrient deficiencies. In addition to the direct loss of crop producing capacity, soil erosion increases the destructiveness of floods and decreases the storage capacity of water in reservoirs.

It is therefore essential that the soils should not be allowed to wash or blow-away more rapidly than they can be regenerated, their fertility should not be exhausted and their physical structure should remain suited to continued production of desired plant materials. Protection of land from further degradation, adoption of various conservation measures, including reclamation and scientific manage-ment of available land stock is very important for a country like India to achieve higher productivity of food, fodder, fuel and industrial raw materials on a substantial basis. Besides, demand for land for providing social priorities such as shelter, roads, industrial activities is increasing at a very fast rate with the rise in population and very often good agricultural and forest lands are being diverted to such use. It is, therefore, necessary to keep soil in place and in a state favourable to its highest productive capacity.

Soil Erosion

The process of destruction of soil and the removal of the destroyed soil material constitute soil erosion. According to Dr. Bennett "the vastly accelerated process of soil removal brought about by the human interference, with the normal disequilibrium between soil building and soil removal is designated as soil erosion".

Types of Soil-Erosion

Erosion of soil by water is quite significant and takes place chiefly in two ways (a) Sheet erosion, (b) Gully erosion.

Sheet Erosion

Sheet movement of water causes sheet erosion and depends on the velocity and quantity of pronounced surface runoff and the erodability of the soil itself. In such cases, the soil is eroded as layers from the hill slopes, sometimes slowly and in-sidiously and sometimes more rapidly. Sheet erosion is more or less universal on:-

- all bare follow land,
- all uncultivated land whose plant cover has been thinned out by over grazing, fire or other misuse, and
- all sloping cultivated fields and on sloping forest, scrub jungles where natural porosity of soil has been removed by heavy grazing, felling of trees or burning etc.

The particles loosened and shifted by the rain drops are carried down slope by a very thin sheet of water which moves along the surface. The impacts of the raindrops increases the turbulance and transporting capacity of this unchannelized sheetwash which results in the uniform skimming of the top soil. Sheet erosion is considered as dangerous as it may continue for years but may or may not leave any trace of the damage. Sheet erosion is common in the Himalayan foot-hills, in Assam, Western ghats and Eastern ghats. When sheet erosion continues unchecked, the silt laden run-off forms well-defined minute finger shaped grooves over the entire field. Such thin channeling is known as 'rill-erosion', which is active over wide areas in Bihar, Uttar Pradesh, Madhva Pradesh and in semiarid areas of Maharashtra, Karnataka, Andhra Pradesh and Tamil Nadu.

Gully Erosion

On a gentle slope, adequately covered by vege-tation, clay soil will resist erosion to a great extent and the water forms small rivulets which can then erode deeper. The rivulets in turn join together to form larger channels until gullies are formed gradually deep gullies cut into the soil and then spread and grow until all the soil is removal from the sloping ground. This phenomenon once started and if not checked, goes on extending and ultimately the whole land is converted into a bad-land topography. Gully erosion is more common in areas where the river system has cut down into elevated plateaus so that feeders and branches carve out an intricate pattern of gullies.

Apart from this, it also takes place in relatively level country whenever large blocks of cultivation give rise to con-centration of field run-off.

Wind Erosion

It occurs in dry climatic areas having a sparse and low vegetation cover on mechanically weathered, loosened surficial material. Dust storms are the principal agents of wind erosion. The top soil is often blown off from the surface rendering it infertile. Besides, with the decrease in the wind velocity coarse sand particles get deposited in some areas covering the existing soil and rendering it unproductive.

SOIL BIOLOGY AND THE BIOLOGICAL MICRO-ENVIRONMENT

Organic matter, microscopic and macroscopic organisms (*e.g.*, fungal hyphae and invertebrates), detritus from fungi and animals, and bacteria, and biological exudates, all assist in stabilizing soil structure. The role of each part of the biomass differs according to its size. Broadly, large aggregates greater than 250 m m diameter (macro-aggregates), are stabilized by their inherent physical structure, wetting and drying cycles, and organic matter. Micro-aggregates (< 250 m m) are stabilized by live or dead roots, fungi, invertebrates and microorganisms.

The populations of soil organisms of all sizes are linked functionally through their roles in the degradation of various forms of organic material. The latter includes live and dead plant material and other live or dead organisms. This shows that animals such as nematodes and some fungi feed directly on live plants while other fungi and bacteria feed predominantly on litter. Earthworms and other large invertebrates create, and inhabit, burrows and pores, and are very mobile. The most notable of these are termites, which are divided into three groups according to the structure of their nests: those that build mounds (a) above ground, (b) on the soil surface, and (c) below ground. Small arthropods, microfauna and fungi live mostly in larger voids and in association with roots. Foster (1988) reviewed the location of the various types of soil-dwelling organisms and found that fungi, which constitute about 80% of the biomass in many soils, tend to be restricted to the rhizosphere of roots, to larger pores between aggregates and to the surface of aggregates. Bacteria, by contrast, are found on roots in the rhizosphere, in small colonies in the larger micropores, within aggregates and on and within cell debris.

Organic Matter

Both plants and animals provide inputs of organic matter to soils. Once within the soil organic residues can be distinguished on the basis of their chemical structure (*e.g.*, old lignified humic substances that degrade slowly), by their source (plant or animal) or by location.

The standing crop of litter in semi-arid grasslands is usually more than 3 t/ha and in temperate dry steppe may exceed 11 t/ha. There has been much debate about the relative contents of organic matter in tropical and temperate soils. Within those wet-and-dry climates that have hot summers assisting rapid decomposition, there is no evidence of inherently lower levels of organic matter in the tropics than in comparable temperate regions. Kowal and Kassam (1978) and Juo and Payne (1993) review the role of organic matter in tropical soils. Here, it is sufficient simply to state that organic matter has various interrelated effects on soil fertility. In particular it should be noted that both chemical and physical effects are of relatively great importance in the soils of the semi-arid tropics because these generally have low cation exchange capacity. The relative importance of litter (crop residue) and manure as inputs of organic matter, varies between cropping systems and spatially within a system. Here, most of the above-ground crop residue is fed to animals, but an equal amount of below-ground crop material enters the soil organic matter pool.

Where alley cropping and agroforestry are practised, values are more variable, but possible inputs could be very significant where the trees, from which the litter is taken, are grown away from the annual crops. If, say, two-thirds of the leaves from leguminous trees are harvested annually, litter values will be substantially higher and the material of better quality than the leaf and stem residue from an annual crop likely to be recycled in the field. Some qualifications, however, should be made. Tree root material is not available for decomposition in the crop field unless it is spatially overlapping (*e.g.* as an intercrop), in which case the trees will compete with the crop for soil nutrients, water, light and space.

The proportion of animal and human manure used on cropland is more variable. Some farmers have developed stable systems which strongly emphasize the use of animal manure on crops. For example, Norman *et al.*

(1982) describe how farmers in northern Nigeria managed to apply 4 t/ha of manure to their heavily-cropped crop land though they had only 3 cows each. Many other farmers do not ensure adequate recycling, either because they are more concerned with livestock management or they do not know the importance of maintaining a 'zero nutrient budget' to replace nutrients removed by the crop. For example, Norman *et al.* (1982) also describe farmers with 10 cows each, who applied only 1.9 t manure/ha to their crops.

Within a cropping system, manuring practice varies with location. There is transference towards the centre of the system. On traditional farms, the area near the household or village is highly fertilized with human and animal manure while more distant fields receive little or no organic matter. Fussell (1992) describes such a traditional 'ring' farming system in semi-arid west Africa. Here, if houses are thatched, the village needs rebuilding or moving every 2 to 4 years. Moving takes advantage of the fertility gradient. Where the huts are not moved, the fertility gradient becomes steeper with time. Rather than trying to even out fertility by labour-intensive transport of manure, farmers vary the cropping of the fields. Continuous cropping of millet is sustainable close to the hut or village where there is plenty of human and animal manure but crop rotations are essential at the periphery.

MANAGEMENT FOR MAINTENANCE OF SOIL BIOLOGY AND NUTRITION

The aim of management should be to create balanced organic matter and mineral budgets. It should ensure that, over several years (a complete crop rotation), soil organic matter is not depleted and that nutrients added equal or exceed those removed by cropping or lost in various ways.

When managing organic matter farmers should recognize that the effects of animal and human manure, sewage sludge and plant residues last longer than those of green manure crops. Green manures, though valuable, usually last only one or two seasons because they are incorporated before they are mature and lignified.

Juo and Payne (1993) advise: 'In spite of the many proven benefits of soil organic matter, its management and recycling in an intensified, modem agro-ecosystem must necessarily revolve around two fundamental characteristics of the system, namely, the availability of organic material at the farm level, and the economic incentive for conserving and recycling organic matter'. To this may be added consideration of the benefits and costs of treating and transporting human wastes from cities. On-site organic matter, organic material brought in from outside and industrially-produced fertilizers will be used according to their benefit cost ratios and the attitudes of governments and crop managers. Tillage reduces the frequency of VAM, at least in invertebrates. Tillage reduces earthworm populations 10-30-fold, both killing them directly and destroying their burrows. Haines and Uren (1990) show that differences in earthworm populations between tillage treatments and stubble burning are reflected by 50% differences in numbers of soil pores.

Tillage also effects the location, numbers and activity of microorganisms within the soil, depending on the type of tillage. For example a mouldboard plough that inverts topsoil has more effect than minimal tillage. Doran (1987) found microbial biomass and potentially-mineralizable nitrogen to be 54% and 37% higher respectively in the top 7.5 cm of no-till than ploughed soil.

These would be critical differences if translated to the semi-arid tropics. Microbial and fungal biomasses may be 20-70% higher after 1 and 2 years no-till. However, conventional tillage, by burying topsoil, may cause microbial populations and their activities to be higher than under no-till at depths of say 7.5 to 15 cm. Differences in soil organisms (*e.g.*, microbial biomass) under differing tillage treatments are seasonal, generally being greatest in the period when the weather is most favourable for soil organism activity and plant growth. Rasmussen and Collins (1991) summarize many data on the impact of tillage on soil properties and conclude that stubble-mulching with zero tillage conserves up to 2% more organic matter per year than ploughing.

SOIL COMPACTION AND BUILDING

Soil compaction is defined as the method of mechanically increasing the density of soil. In construction, this is a significant part of the building process. If performed improperly, settlement of the soil could occur and result in unnecessary maintenance costs or structure failure. Almost all types of building sites and construction projects utilize mechanical compaction techniques

Soil density is used almost exclusively by the transportation industry to specify, estimate, measure, and control soil compaction. Soil density can be easily determined via weight and volume measurements. The objective of compaction is to stabilize soils and improve their engineering properties. Source: US Dept. of Transportation

There are five principle reasons to compact soil:-

- Increases load-bearing capacity-
- Prevents soil settlement and frost damage-
- Provides stability-
- Reduces water seepage, swelling and contraction-
- Reduces settling of soil-source Soil Compaction Handbook.

POROSITY

Bulk Density is an indirect measure of soil pore space. In fact, there is a formula: 100-((B.D. ÷ Particle Density) X 100)=% pore space or POROSITY. Under field conditions, pore spaces are occupied at all times by air and water. "Tortuous pathways" best describes soil pores. Porosity, when expressed as a percent, is the same thing as percent pore space. Soil particles have irregular shapes, and thus the spaces or pores between them vary irregularly in size, shape and direction. Sandy soils have large continuous pores, while clays have small pores which transmit water slowly. Clays, however, contain more pore space than sandy soils, because of the pores inside of the soil peds. To growing plants, pore sizes are of more importance than total pore space.

Particle Density

In the laboratory you will determine porosity and particle density for the sandy soil used previously. Particle density will also be determined for this soil. Particle density only takes into account the volume occupied by the solid particles. It excludes the volume occupied by air and water. Since a large portion of most soils is composed of particles derived from quartz minerals, the particle density of most soils is near 2.6 g/cc, which is the density of quartz. Variations in the particle density are due to the presence of heavier minerals like iron oxides or lighter organic components.

Significance of Porosity

Above, a weakly developed soil is on the left and a well developed soil is on the right. Where clay accumulation is minimal, the macro pore space is about the same with increasing depth. However, where clay has moved into pores, the amount of macro pores and the rate of water movement both decrease. Soils with clay films will have slow water flow through the Bt horizon.

Some soils are difficult to aerate by plowing due to the turf cover. For lawns, playgrounds, or football fields, aeration is accomplished by making small holes in the turf with solid tines or using hollow tines to pull out a small core. Later the holes can filled with sand to promote air exchange. This is called *aerifying* and *topdressing*.

Many home lawns are compacted because of the construction activities when the home was built or due to people walking or riding on the turf when the soil is wet. Aerification will improve the turf growth on these lawns.

AERATION EQUIPMENT=TINES & JOHN DEERE

Soil Temperature Regimes

The temperature of a given soil at a given time is dependent upon the gains and losses of heat energy. Generally, dark surfaces will absorb more heat than light surfaces. However, the amount of water in the soil is an important factor. Losses of the absorbed heat are by radiation back into the atmosphere as long-wave radiation, heating the air and cooling the soil.

These soils warm quickly in the spring due to their dark surface.

Soil temperature regimes are used to classify soils. They are defined according to the average annual soil temperature in the root zone. The use of soils for agriculture and forestry is closely related to soil temperature, due to the specific requirements of plants. Over most of the earth, daily soil temperatures below 50 cm deep seldom change. To approximate the mean annual soil temperature, 2 º F is added to the mean annual air temperature. If we do this for the annual temperature map of Minnesota and follow the 45ºF line, which would be 47º soil temperature, the state is divided along a line from the Twin Cities to the South Dakota border, and soils are frigid to the north and mesic to the south.

SOIL BULK DENSITY DETERMINATION

Soil weight is referred to as soil bulk density. Density is the mass of material contained within a given volume. This is the idea that a given size of box may be heavy or light, depending upon what kind of material it contains. If the box were filled with wood, it would be light when compared to having it filled with lead. The weight of water is the reference for density measurements: 1 gram of water=1 cubic centimeter (cc), & 1 cc water=1 ml. & 1 cubic foot of water=62.4 lbs.

The bulk density of a soil is the mass of dry soil per unit of "bulk" (total volume of soil or soil particles & pore space).

B.D. = mass of oven dry soil (grams) ÷ total volume of soil (cm^3)=grams/ cc

The bulk density takes into account the total soil volume (the space occupied by the solid particles plus the space occupied by the air of the pores or pore space). To determine the mass of the soil—since air does not have any significant weight—we can just weigh the oven dry soil on a balance. The volume of the soil can be determined by pouring the soil into a graduated cylinder and measuring the volume that it occupies.

The problem with this procedure is that structural aggregates may be crushed or compacted once they are removed from the soil and placed into the cylinder. A better procedure is one in which the aggregates could be removed from the soil and frozen exactly the way they were in the soil. Plastic fixatives enable us to do this.

A large aggregate has been removed from the soil and is being fixed by dipping it into a saran solution. This will make the clod impervious to water. The weight of our clump of soil can be obtained by hanging it on the balance or placing it on the metal weighing tray.

The volume of the clod also needs to be determined. This can be obtained by weighing the sample again, only this time the sample will be in water. The weight of the sample now will be less, since the sample will be buoyed up by the amount of water it is displacing.

Archimedes most famous theorem gives the weight of a body immersed in a liquid, called Archimedes' principal. If you want to know more about the mathematician who determined that this procedure would work, go to Archimedes.

Thus, by subtracting the weight of the clod in water from its weight in air, we obtain the weight of the water displaced by the clod, which equals the volume of water displaced by the clod (because 1 cm^3 of water=1 gm. of water) and is equal to the volume of the soil clod.

Clod Method of BD Determination =Mass of Clod ÷ Volume of water displaced

Volume of water displaced=weight of water or the (gms clod in air)-(gms clod in water)=(grams or volume of water displaced)

Cylinder method of BD Determination=Mass of oven dry soil (gms) ÷ total volume of soil (cm^3)

Another method of determining bulk density is by using soil cores. These are obtained with a probe that forces a cylinder into the soil. The soil can be removed from this cylinder and cut into sections and placed in another cylinder for transportation back to the laboratory. Care needs to be taken to avoid compacting the soil during and after obtaining the core.

BD of soil (using core method)={(mass of soil + mass of core)-(mass of core} ÷ (volume of core)

volume of core=Pi (3.1416) $r^2 \times h$

r=radius of core; h=the height of core.

For the cores in the lab, r=2.45cm (½ of Diameter); h=5 cm

$V=3.1416 \times 2.45^2 \times 5=94.3\ cm^3$;

Bulk density is a fairly easy determination that will yield information about the soil that will be significant for determining the soil's potential for plant growth or a building foundation. Remember it is the oven dry mass/ volume soil. or g/cc.

SIGNIFICANCE OF BULK DENSITY

The bulk density of the soil will play an important role in determining if the soil has the physical characteristics necessary for plant growth, building foundations or other uses. From the laboratory investigations you will obtain very high bulk densities for the Bt and sandy soil. The weight of the soil is important if you are going to be lifting it or hauling it long distances. One of the main reason for sod farms to be on organic soils is to reduce the cost of transportation because the organic soil is so much lighter than mineral soils. The organic sod is easier to handle. Most sod farms in Minnesota are on the peat soils north of the Twin Cities.

Sometimes we are interested in the weight of an acre of soil for erosion comparison purposes. Erosion of 5 tons per acre sounds like a lot. However the weight of an acre furrow slice on average is 2,000,000 lbs., so 5 tons per acre seems like a small amount. Five tons per acre is only about 34 thousandths of an inch thick (.034 inches). However, many areas have lost 10 to 100 times that amount. (Note: an acre furrow slice, or AFS, is a three-dimensional volume of soil which is one acre in area and 7 inches deep.)

Weight per AFS=BD × Volume

Example: Volume of soil=Soil depth × 43560 ft.^{2} (area of 1 acre)

For conversion of g/cc to lbs/ft^3, multiply BD × 62.4 lbs/ft^3 (weight of ft^3 water)

How many tons are lost if a soil with a B.D. of 1.2 g./ cm^3 erodes 5 inches per acre?

answer: (1.2 × 62.4)=(weight in lbs/ft^3) X Volume: (where Volume=(5 in. ÷12 in/ft) × 43560 ft)^2 or (74.88 lbs/ft^3) × 18,150 ft^3=1.3 million lbs or divide by 2000 lbs / ton or weight=679 tons.

Soil Factors Impacting Tillage

Another factor similar to carrying soil is its potential to be moved short distances, such as in plowingor rototilling. The soil that is heavier will be more difficult to move. However, another factor is the ability of the soil to stick together, or the soil's consistence. Clay soil is noted for its stickiness and large energy requirements in tillage. Farmers refer to it as "heavy," but they really mean it is difficult to plow, not that it has a high bulk density; clay soils generally will have a lower bulk density than sandy soils (clay=1.3 g/cc vs sand=1.6g/cc). Sandy soils have a higher bulk density, but are easier to plow since they have weaker consistence. Thus they are often referred to as "light soils." The lower B.D. of clays is due to their better aggregation.

Rototilling the soil reduces the bulk density by "fluffing" the soil. For a look at tillage implements go to Tillage Implements

Soil Consistence

Soil consistence is the soil's ability to cohere or stick together. The soil's consistence may be evaluated at three moisture conditions: air dry, moist, and wet. Moist consistence is evaluated by placing the soil between the thumb and forefinger and gently applying pressure. The ease with which a ped can be crushed determines the consistency.

Terms commonly used to describe moist consistence are:

Loose-Non-coherent when dry or moist; does not hold together in a mass.

Friable-When moist, crushes easily under gentle pressure between thumb and forefinger and can be pressed together into a lump.

Firm-When moist, crushes under moderate pressure between thumb and forefinger, but resistance is distinctly noticeable.

Plastic-When wet, readily deformed by moderate pressure but can be pressed into a lump; will form a "wire" when rolled between thumb and forefinger.

Sticky-When wet, adheres to other material and tends to stretch somewhat and pull apart rather than to pull free from other material.

Hard-When dry, moderately resistant to pressure; can be broken with difficulty between thumb and forefinger.

Soft-When dry, breaks into powder or individual grains under very slight pressure.

Cemented-Hard; little affected by moistening.

In air dry conditions, the resistance to rupturing when rubbed is measured. At intermediate moisture content, the soil's resistance to shearing forces by thumb and finger is noted. In the wet condition, its plasticity—ability to be molded and stickiness—are measured.

One significance of the B.D. of a soil is the amount of surface area a soil has. The more surface area, the more ability to retain water and nutrients. Notice that the two soils pictured below weigh the same, but are significantly different in surface area. Which soil has the higher B.D.?

COMPACTION, POROSITY AND SOIL TEMPERATURE

Soil Compaction

Soil compaction works something like the soil particles seen below. When the soil is randomly fluffed, as on the left, the amount of pore space will be 45 to 55%. When we push the soil particles closer together, we remove the pore space, and the bulk density will increase. Compaction increases the bulk density by reducing the pore space.

Compaction can also result in the changing of the proportion of pore sizes. Notice in the compaction diagram that compacted soils not only have a lower total pore space (resulting in higher bulk density), but also less macro pores and more micro pores.

This will often result in excess water retention.

When we walk on the soil we compact it. Eventually the compaction prevents the growth of plants. This trail TRAIL behind the Soils Building shows the effect of compaction and samples from the soil will be used in the lab to determine the amount of compaction.

Compaction changes the amount and size of pores.

Deep plowing can reduce compaction below the plow zone.

C.O.L.E.

One of the important engineering properties of soils is related to the clay content and consistency and is called the coefficient of linear extensibility or C.O.L.E. It is a measure of the shrink-swell potential, or the volume change of the soil with changes in moisture content. If the C.O.L.E. exceeds.09, significant shrink-swell potential can be expected.

Soils with high C.O.L.E. values can cave in basements, displace walls, and break gas or water pipes. Knowing which soils have high C.O.L.E. values is an important step in selecting sites for construction activities. Compacting soils reduces the ability of soils to take on water and reduces their shrink/swell potential.

C.O.L.E.=(Length Moist ÷ Length Dry)-1

SOIL ANALYSIS: A KEY TO SOIL NUTRIENT MANGEMENT

High yields of top-quality crops require an abundant supply of 16 essential nutrient elements. In addition to providing a place for crops to grow, soil is the source for most of the essential nutrients required by the crop. Our soil resource can be compared to a bank where continued withdrawal without repayment cannot continue indefinitely. As nutrients are removed by one crop and not replaced for subsequent crop production, yields will decrease accordingly. Accurate accounting of nutrient removal and replacement, crop production statistics, and soil analysis results will help the producer manage fertilizer applications.

A soil analysis is used to determine the level of nutrients found in a soil sample. As such, it can only be as accurate as the sample taken in a particular field.

The results of a soil analysis provide the agricultural producer with an estimate of the amount of fertilizer nutrients needed to supplement those in the soil. Applying the appropriate type and amount of needed fertilizer will give the agricultural a more reasonable chance to obtain the desired crop yield.

OBJECTIVES OF SOIL ANALYSIS

- To provide an index of nutrient availability or supply in a given soil. The soil extract is designed to evaluate a portion of the nutrients from the same "pool" used by the plant.
- To predict the probability of obtaining a profitable response to fertilizer application. Low analysis soils may not always respond to fertilizer applications due to other limiting factors. However, the probability of a response is greater than on a high analysis soil.
- To provide a basis for fertilizer recommendations for a given crop.
- To evaluate the fertility status of the soil and plan a nutrient management programme.

Chemical analysis of plant composition indicates chemicals or elements present in a crop at maturity or when it is harvested. For example, 1,250 lb of lint cotton contains approximately 125 lb of nitrogen (N), 20 lb of phosphorus (P), and 75 lb of potassium (K).

The essential question in fertilization is, "How much nutrient must be added to the soil as fertilizer for a given amount to be taken up by the growing plant?" The crop utilizes only a portion of the available nutrients in the soil. This means that more nutrients must be present than are removed by the crop. The amount added varies according to the level already present in the soil and the crop's need for the nutrient involved. The soil analysis is the starting point, since it measures the level or content presently in the soil.

The soil analysis along with the information provided in the information sheet, is interpreted and reported in terms of the nutrients needed to supplement those in the soil. With this information, producers can add sufficient nutrients for the correct balance to obtain high yields.

Limiting Factors

Crop yields are determined by a variety of factors including crop variety selection, available moisture, soil fertility, crop adaptation to the area, and the presence of diseases, insects, and weeds. The soil analysis and its interpretation deal only with the fertility level (plant nutrients) of the soil. Recommended fertilizer will provide sufficient nutrients for the best possible yields. Other factors of production or management may still cause low yields, even though nutrients are adequate.

Carryover

If yields are only partial in relation to a large amount of fertilizer applied, many of the nutrients are carried over for use by the next crop. It is this carryover, or residual effect, from one year to the next that makes heavy fertilizer applications practical in the face of other limits to yield.

Yields to Expect

A certain fertilizer application cannot be expected to produce a specific yield such as two bales of cotton or nine tons of hay. It is more realistic to assume that a balanced fertilizer programme assures that nutrients are not the limiting factor in yields obtained. Research has shown that producers who use a balanced fertilizer programme obtain consistently better yields than those who don't.

The Soil Analysis Report

After the soil is analyzed, fertility recommendations are made based on amounts of actual nutrients in the soil, not on the amount of any particular fertilizer or mixture. For example, if 100 lb of N were recommended, that amount could be supplied by approximately 300 lb of ammonium nitrate (33%N), 220 lb of urea (45%N), or 120 lb of anhydrous ammonia (82%N). Likewise, a recommendation of 60 lb of P205 per acre could be added as 133 lb of 45% triple superphosphate.

Fertilizer Labeling

Nitrogen is expressed on the elemental basis as "total nitrogen" (N). Phosphorus is expressed on the oxide basis as "available phosphoric acid" (P205). Potassium is expressed as "soluble potash" or potassium oxide (K20).

In reality, there is no P205 or K20 in fertilizers. Phosphorus exists most commonly as monocalcium phosphate, but also occurs as other calcium or ammonium phosphates. Potassium is ordinarily in the form of potassium chloride or sulfate. Furthermore, P205 and K20 are not absorbed by plants.

Plant roots absorb most of their phosphorus in the form of orthophosphate ions, H2P04-, and most of their potassium as potassium ions, K+. For these reasons, the elemental expression (N-P-K) is used in all of the recent research publications. Conversions from one form of P and K to another can be made using the following formulas.

%P=%P205 x0.437 %K=%K20x0.826

%P205=%Px2.29 %K20=%Kx1.21

INTERPRETATION OF THE SOIL ANALYSIS REPORT

The soil analysis report contains two parts: characterization and fertility status of the soil, and fertility recommendations. Soil characterization (pH, texture, percent exchangeable sodium, percent organic matter, and salinity expressed as electrical conductivity) is explained in the report. The fertility status is reported as nutrients available to the plant. The second part, fertility recommendation, contains the suggested amounts of fertilizer to apply. These amounts are based on the crop requirements, management practices affecting the crop (as shown in the information sheet), the present fertility level of the soil, and the yield goal desired by the producer. Special notification is given if the tests indicate that a salt or sodium hazard exists or if the information provided shows any other specific problems.

Soil amendments or treatments to reduce a sodium or salt hazard will be recommended if requested. In general, application of gypsum is suggested for reducing a sodium hazard, and leaching is recommended in most cases to lower salt content in the soil. Gypsum or leaching requirements are calculated and reported if requested.

Where to Get Soil Analyzed

There are many soil testing laboratories in New Mexico, Texas, Colorado, and Arizona. Basic soil testing packages vary in price and number of analyses. Many labs are participating in the Western Region Soil Testing Proficiency Programme. Programme participants share identical soils and compare results quarterly. This process assures the clients that the lab is striving for consistency and accuracy in lab analyses. Recommendations will undoubtedly vary from lab to lab. Often the best recommendation will come from the local Extension service. The choice of labs is at the client's discretion but should be based on report readability, result accuracy, turn-around time, and cost factors. New Mexico specialists can assist with many questions regarding plant health. Remember, a soil analysis is only as good as the soil sample taken.

SOIL ANALYSIS: KEY TO NUTRIENT MANAGEMENT PLANNING

Soil provides a reservoir of nutrients required by crops and also therefore for animals but not necessarily at optimum levels of immediate availability to plants. The purpose of soil analysis is to assess the adequacy, surplus or deficiency of available nutrients for crop growth and to monitor change brought about by farming practices.

This information is needed for optimum production, to avoid transferring undesirable levels of some nutrients into the environment and to ensure a suitable nutrient content in crop products.

Farm assurance schemes, buyer's protocols and codes of practice are increasingly demanding more accurate fertiliser recommendations which must depend on the nutrient-supplying capacity of the soil. Regular soil analysis should be undertaken as a vital part of good management practice.

SOIL CONSERVATION

Soil conservation is a set of management strategies for prevention of soil being eroded from the Earth's surface or becoming chemically altered by overuse, acidification, salinization or other chemical soil contamination.

It is a component of environmental soil science. Decisions regarding appropriate crop rotation, cover crops, and planted windbreaks are central to the ability of surface soils to retain their integrity, both with respect to erosive forces and chemical change from nutrient depletion.

Crop rotation is simply the conventional alternation of crops on a given field, so that nutrient depletion is avoided from repetitive chemical uptake/ deposition of single crop growth.

EROSION PREVENTION

Practices

There are also conventional practices that farmers have invoked for centuries. These fall into two main categories: contour farming and terracing, standard methods recommended by the U.S. Natural Resources Conservation Service, whose Code 330 is the common standard.

Contour farming was practiced by the ancient Phoenicians, and is known to be effective for slopes between two and ten percent. Contour plowing can increase crop yields from 10 to 50 percent, partially as a result from greater soil retention.

There are many erosion control methods that can be used such as conservation tillage systems and crop rotation. Keyline design is an enhancement of contour farming, where the total watershed properties are taken into account in forming the contour lines. Terracing is the practice of creating benches or nearly level layers on a hillside setting. Terraced farming is more common on small farms and in underdeveloped countries, since mechanized equipment is difficult to deploy in this setting. Human overpopulation is leading to destruction of tropical forests due to widening practices of slash-and-burn and other methods of subsistence farming necessitated by famines in lesser developed countries. A sequel to the deforestation is typically large scale erosion, loss of soil nutrients and sometimes total desertification.

Perimeter Runoff Control

Trees, shrubs and groundcovers are also effective perimeter treatment for soil erosion prevention, by insuring any surface flows are impeded. A special form of this perimeter or inter-row treatment is the use of a "grassway" that both channels and dissipates runoff through surface friction, impeding surface runoff, and encouraging infiltration of the slowed surface water.

11

Custodians of the Forest

INTRODUCTION

Isang Bagsak South-East Asia is an enabling programme in more ways than one. The implementation of this networking programme, through its basic strategy of participatory development communication (PDC), is giving rise to experiences and insights on involving the community in development work through communication. More significantly, this involvement hinges on the use of information and communication technology – a 21st-century tool whose applications for development are currently gaining ground in developing countries.

The programme is being implemented in the Philippines by the College of Development Communication (CDC) of the University of the Philippines at Los Baños and its partner in the first cycle, Tanggapang Panligal Para sa Katutubong Pilipino, or the Legal Assistance Centre for Indigenous Filipinos (PANLIPI). The decision regarding the Philippine partner for the Isang Bagsak learning network was not an easy one to make. There were many mainstream organizations with vast experience in natural resource management. Resource constraints also figured prominently on the choice of partner.

In the end, the CDC posed one question that made all the difference: who would benefit most from the Isang Bagsak experience? The following rhetorical question also tipped the balance in favour of PANLIPI: why not share the learning from Isang Bagsak directly with the indigenous peoples, the custodians of Philippine forests?

According to the Indigenous Peoples' Rights Act of the Philippines, indigenous peoples refer to a group of people or homogeneous societies identified by self-ascription or ascription by others who have continuously lived as organized communities in a communally bounded and defined territory and who have, under claims of ownership, since time immemorial, occupied, possessed and utilized such territories sharing common bonds of language, customs, tradition and other distinctive cultural traits. Indigenous peoples have, through political, social and cultural inroads of non-indigenous religions and cultures, become culturally differentiated from the majority of Filipinos. Indigenous peoples will include people who are regarded as indigenous on account of their descent from populations who occupied the country before colonization and who retain some or all of their own socio-cultural or political institutions.

A SHOWCASE OF SYNERGY

PANLIPI is an organization of lawyers and advocates of indigenous peoples' concerns. Established in 1985, PANLIPI primarily aims to assist the indigenous peoples in their struggle for the recognition of their rights to their ancestral domains, culture and traditions, and other basic rights. The end goal is to empower the indigenous peoples to the fullest so that they can actively participate in every aspect of Philippine society. PANLIPI's mandate includes the provision of development legal assistance; legal education and outreach; institutional capacity-building; ancestral domains delineation; and resource management planning.

The networking among Isang Bagsak, CDC and PANLIPI may be viewed as a synergistic one right from the start.

From PANLIPI's viewpoint, the promotion of the rights of indigenous peoples to their ancestral domains and the issue of natural resource management go hand in hand. The indigenous peoples are, after all, the best caretakers or stewards of the natural resources mainly because their socio-economic system is sustainable and not destructive.

For its part, CDC found in PANLIPI a partner who could best make use of the very tenets of development communication – that is, the use of communication towards improved socio-economic growth of a community that makes for social equity and the larger unfolding of individual potential. Through Isang Bagsak, the indigenous peoples of the Philippines are given their window to the world of learning and are sharing their own knowledge, insights and experiences with members of the network from various parts of the globe. Likewise, the other members of the network are allowed access to the rich heritage of peoples hitherto unwired to the Isang Bagsak network.

LESSONS LEARNED

Working with PANLIPI is, indeed, a work in progress that is bringing forth a lot of valuable insights for researchers and development workers in general. The introductory stage of the partnership was already fraught with lessons, foremost of which were the ethical considerations. For one, is it ethical to bring in the indigenous peoples to a programme that is quite alien to them in terms of the tools (information technology) to be used? Second, will they agree to become part of the undertaking, and how do we engage their participation?

MUTUAL RESPECT AND TRUST

There was conscious effort on the part of the CDC to show respect and to gain the indigenous peoples' trust at the start of the project. Before any agreement was inked, the project team, with PANLIPI's help, made a point of observing two things:

- understanding the cultural context; and
- securing the indigenous peoples' free and prior informed consent.

Henceforth, a series of consultations with the indigenous peoples who would be involved in the learning network was undertaken. The CDC team, together with the PANLIPI lawyers, literally sailed, crossed rivers and hiked through mountains to dialogue with the elders of the cultural communities. This was part of securing their free and prior informed consent, in keeping with the PANLIPI protocol and the nature of participatory development communication.

These are two critically important values that many development workers seem to be taking for granted, but which the Isang Bagsak experience is bringing to the fore in this undertaking.

Genuine dialogue between and among the CDC, PANLIPI and the indigenous peoples paved the way for the participation of people from four areas: Mangyans from Mindoro; Tagbanuas from Palawan; Kankanaeys from the Cordilleras; and Aetas from Zambales. Iterative consultations leading to consensus-building were held. First, the CDC team conferred with the PANLIPI staff and lawyers assigned to the areas. Once PANLIPI's cooperation was ensured, dialogues were held with indigenous elders and leaders, and eventually with the members of the community.

The series of dialogues highlighted the following points:

- Knowledge sharing: Isang Bagsak offered an opportunity for the PANLIPI network and the indigenous peoples in the four areas to not only learn from others and from one another about natural resource management and PDC, but to contribute their own knowledge to the network.
- Assurance that the indigenous knowledge systems (IKS) would be respected: the indigenous peoples voiced a strong concern about the possible piracy of their traditional knowledge, as was their experience in the past. They were wary of outsiders 'stealing' their indigenous knowledge in the guise of development work. Prior experience of being exploited by researchers from outside (foreign and local) was a big stumbling block to securing the indigenous peoples' free and prior informed consent. Team members took pains to explain that nothing that did not come from the people, through their representatives in the programme, would be posted on the website. This was, in essence, a manifestation of respect for intellectual property rights.

RELEVANCE TO INDIGENOUS PEOPLES' CONCERNS

The indigenous peoples' decision to participate in Isang Bagsak rested largely on the realization that the skills they would develop would be useful to the concerns that they were advocating, such as the Ancestral Domains Sustainable Development and Protection Plan and other important provisions of the newly passed Indigenous Peoples' Rights Act. PANLIPI participants saw the Isang Bagsak experience as an excellent opportunity to interact with professionals from other countries. Through the sharing of experiences, one important insight surfaced: the affirmation that all along, PANLIPI and the indigenous peoples have been practising the principles of development communication, in general, and those of PDC, in particular. The training that the PANLIPI participants have undergone in the course of their participation in Isang Bagsak has also afforded them the opportunity to write and communicate their ideas clearly.

INFORMATION TECHNOLOGY FOR INDIGENOUS PEOPLES

Technology could be an awesome development for indigenous peoples, only very few of whom have been initially exposed to computers, much less to the internet. Part of their initial reluctance to be part of Isang Bagsak was apprehension about the technology itself. Connectivity/accessibility also loomed as a potential problem.

To address these concerns, PANLIPI, for its part, has begun sponsoring a series of training sessions on information technology for the indigenous leaders involved in Isang Bagsak. CDC's information technology team trained Iraya Mangyan leaders in Mindoro during February 2004. A user-friendly sourcebook on using the computer and the internet was developed by the CDC especially for this training. The sourcebook, in turn, would be useful in future information technology training for indigenous peoples. In addition, PANLIPI has been very supportive in enabling the indigenous peoples to have access to computers and the internet.

CDC and PANLIPI team members agree that one of Isang Bagsak's 'byproducts' is the gradual mainstreaming of the indigenous peoples into the digital age.

PDC THROUGH THE NET: TEAMWORK AT ITS BEST

By the time the project was in full swing, the PANLIPI–Isang Bagsak network had 45 members, 9 each from the 5 indigenous groups. A network within a network was created by the partners. As the saying goes: *t*ogether, *e*veryone *a*chieves *m*ore.

The members were selected by the indigenous communities themselves on the basis of their knowledge and skills on natural resource management issues. Other criteria were also included in the list of qualifications. Prospective team members had to be knowledgeable about natural resource management and indigenous knowledge systems and practices. They had to be capable of facilitating community participation. Finally, they were also expected to understand development work and to be knowledgeable and skilled in promoting a rights-based approach to development.

The resulting team was dubbed *Limang Tagupak,* a Mangyan term representing victory. The name also signifies a five-point star, representing the configuration of the five-team network at the PANLIPI level. The star is also symbolic of a guide.

The sharing and learning process on PDC in natural resource management was carried out in the following manner:

- Local teams undergo orientation and training on Isang Bagsak.
- Local teams participate in the forum:
    - o The CDC posts theme questions.
    - o PANLIPI National Capital Region translates theme questions and sends them to local teams.
    - o Local teams discuss the theme questions, formulate answers and send them to PANLIPI National Capital Region.
    - o PANLIPI National Capital Region translates answers and posts reply.
    - o PANLIPI National Capital Region downloads and translates country comments and sends to local teams.
- Local teams discuss country comments and make their own reflections, which they send to PANLIPI National Capital Region.
- PANLIPI National Capital Region translates and posts reflections.
- The CDC resource person synthesizes.
- PANLIPI National Capital Region translates the synthesis and sends it to the local teams.
- Local teams discuss the synthesis and extract the learning.

The process appears to be a tedious one. True enough, a number of problems hampered the implementation of Isang Bagsak at the PANLIPI level. For one, translations into English took up quite some time; hence, delays in posting were inevitable. Delays were also due to the team members' tight work schedules or their preoccupation with economic activities, leaving them less time to participate in the forum. Access to the internet (lack of equipment, telephone lines, internet terminals/connection) constituted the hardware part of the limitations. Financial costs of consultations and regular meetings, as well as internet rental, posed very real problems as well. So did forces of nature, such as bad weather preventing team members from meeting.

PANLIPI's response to the resource-related problems mirrored its commitment to the partnership. PANLIPI put up counterpart funds for training indigenous peoples on basic computer use and internet navigation, as well as hardware for internet accessibility. It forged cooperative undertakings with non-governmental organizations (NGOs) and other friends to boost its investment in equipment and internet access. To facilitate meetings among the team members, PANLIPI made arrangements for Isang Bagsak activities to coincide with regular meetings of indigenous peoples' organizations.

The foregoing are just four of the insights arising from several months' implementation of Isang Bagsak in the Philippines. Many more could be drawn from the researchers' and the indigenous peoples' experiences as they go through the cycle in the remaining months. Isang Bagsak South-East Asia is a work in progress.

NATURAL RESOURCE MANAGERS

Conservation efforts often fail to take into consideration that alternatives have to be available if local populations are to change their natural resource utilization patterns. Along the Cimanuk River in West Java, efforts to prohibit agriculture on the raised riverbanks to curb seasonal flooding irremediably failed until viable economic alternatives were introduced. Through participatory development communication, the dialogue between researchers, decision-makers and small and landless farmers made it possible for sheep raising to flourish as an alternative to the practice of agriculture on the riverbanks that was making villages more vulnerable to annual flooding.

Fifteen years ago, as we were preparing a baseline study to explore the development options that could improve the living conditions of small farmers and landless farm labourers living along the Cimanuk River in the district of Majalengka, West Java, we wondered whether those resource-poor farmers would be able to participate in managing a strip of public land stretched out along the river. We also wondered whether they would be able to derive benefits from such an undertaking.

At that time, the main concern of the local government and of the communities living along the river was flood control. Every rainy season, heavy downpours filled up the river and flooded the villages. To control this annual flooding, the Department of Public Works bought a strip of land along the river that goes through the district's villages and small towns, and raised the riverbanks to a certain height. Hence, the annual flooding was controlled and the communities were freed from the yearly threat of natural disaster.

However, this flood prevention technique created another problem for the small and landless farmers who had been farming on the riversides long before the banks were raised. According to the new government rules and regulations, farming activities were forbidden on the public land along the river due to the risks of destabilization of the land structure and hindrance of the waterways.

In spite of the government rules and regulations, the small and landless farmers pursued their farming activities as usual. On their part, officials of the Department of Public Works continued to enforce the regulations by eradicating the crops that could possibly weaken the raised riverbank's land structure.

Looking for Alternatives

After several years of playing cat and mouse and tired of conflicting with the local communities, the Department of Public Works finally proposed a win–win solution to the farmers. The latter were finally allowed to farm along the riversides, provided that they grew grass at least 1m away from the river's brim and on both sides of the raised bank. However, they were not allowed to grow bananas or any other type of land crops that could destabilize the land structure.

To further motivate the farmers to grow grass, the government officials promised to supply them with sheep. However, this promise did not materialize very well. The farmers were disappointed with the few low-quality sheep delivered to them and the abundant grass was wasted.

At that critical time, a baseline study was undertaken in order to understand the availability of local natural resources such as land, crops, animals and water, as well as the limitations, problems and opportunities associated with their use and development. Moreover, the study also aimed at understanding the traditional systems and at assessing indigenous capacity. The study concluded that there was potential for involving the small farmers in managing the public land located along the river and the irrigation channels, and for deriving benefits from it, provided that a funding body be put in place.

A follow-up pilot study was then undertaken. Essentially, the challenge consisted in helping farmers to solve their problems and benefit from the abundant grass, while ensuring the conservation of the riverbank. After discussing the best-suited approach, our team decided to use participatory development communication. Thus, linkages were established between university researchers, local government, the Department of Public Works, the local livestock extension service, the village authorities and local farm communities.

USING APPROPRIATE COMMUNICATION TOOLS

Prior to the fieldwork, communication materials such as sound-slide shows, posters and even a comic book were prepared and tested with farmers and extension workers. Development communication graduate students were involved in producing and testing these materials, either through their message design course assignments or through their theses research.

Topics such as the social and economic potential of sheep raising in the district of Majalengka were specifically addressed with local decision-makers and livestock extension workers. Other topics concerning various technical aspects of sheep production, marketing and rural family budget management were developed for extension workers, farmers and their wives. The fieldwork started once the communication materials were ready. In the first quarter of 1992, a memorandum of understanding was officially signed between the government of the district of Majalengka and the Faculty of Animal Science at Bogor Agricultural University.

During the ceremony, a sound-slide show regarding the social and economic potentials of sheep raising in the district of Majalengka was shown to the government officials and to the local House of Representatives, prior to signing the agreement. This created enthusiasm among the officials attending the ceremony. Moreover, the local media covered the event, which was broadcast on the evening news by the regional television station. The next step consisted of promoting capacity-building for the local livestock extension personnel and farmer leaders. Once a month, researchers conducted an on-site training meeting for the livestock extension workers and other officials of the district livestock service. The technical aspects of sheep production, including housing, feeding, breeding and sheep healthcare, were addressed. Two sound-slide programmes about sheep production were shown and discussed with the extension workers.

A pre-test and a post-test on the training subjects were always administered to the extension workers prior to and after the slide show in order to build up the agents' interest and to focus their attention on the subjects. Moreover, the tests were also useful in indicating whether the sound-slide shows had been effective in improving the agents' knowledge on the topics that were being discussed.

Following the agents' training, a short meeting was conducted for farmer leaders of the four pilot sites. On that occasion, they were informed that they were entitled to receiving an in-kind loan in the form of sheep. If they agreed, they would have to return a certain number of the sheep offspring to the project for a period of four years.

After the meeting, participants were invited to visit a senior farmer leader in the village of Balida to witness how he collected forage. One month later, they joined an excursion to the district of Garut to visit a sheep-raising initiative. The Garut sheep were particularly well known for their fast growth and prolificacy.

Trying out New Techniques

Upon their return, every farmer leader was asked to build a good sheep house capable of housing two adult rams and ten ewes, and to prepare a plot to grow improved varieties of grass and legumes to feed the sheep for demonstration purposes.

Intensive supervision and backstopping activities were provided to the farmer leaders in the following month to allow them to get first-hand experience in raising the new sheep breed in their own environment. After one month, the farmer leaders had enough experience in taking care of the sheep and were able to provide information about the Garut sheep raising to other farmers in the villages.

Upon completion of the trial run among the farmer leaders, monthly evening training meetings were conducted for the rest of the farmers in the four research sites. The first training meeting was about building a good low-cost and healthy sheep house using local materials, while the second one focused on the appropriate feeding for this specific breed of sheep. During the meetings, all farmers were given pre-tests and post-tests in order to measure their knowledge regarding the training subjects prior to and after the audio-visual presentations.

About two weeks after the training sessions, the farmer leaders were asked to check whether all the farmers had made or had improved their sheep houses for the incoming new sheep. Farmers who had not done so were encouraged to rebuild or improve their sheep houses.

Two weeks after completing the necessary preparations, a farmer leader with extensive experience in sheep breeding was asked to join the researchers to select rams and ewes for 100 farmers in the four research sites.

The training meetings continued after the sheep had been distributed to the farmers. The topics addressed ranged from technical aspects of sheep raising to rural family budget management.

Meanwhile, the research assistant, together with the farmer leaders, visited every farmer once a week to check on the sheep's condition and health and to provide further backstopping activities to the farmers.

About one year after the distribution of the sheep, all farmer leaders were invited to a meeting in the district capital to share their experience in Garut sheep raising with farmers from other villages. This farmer-to-farmer communication turned out to be an excellent medium for raising other farmers' interest in Garut sheep production, better than anything that researchers or extension workers could have done.

Promising Results

The initiative had a first positive result when the head of the local livestock service, using the narrative progress report for the first year, was able to convince the local government to provide his office with extra funding

Towards the end of the second year, results began to be experienced at the grassroots level, when several farmers began to return the sheep offspring to the initiative. According to a previous agreement, the sheep offspring were to be split fifty–fifty. The returned offspring were to be revolved to other farmers interested in sheep raising. This was an indication that the initiative was having a positive impact on the small farmers involved. Indeed, between September and December 1993, the sheep population increased by 159 per cent. One year later, the increase reached 271 per cent, while at the end of

1995, it had risen to 306 per cent. The value of the participating farmers' assets also increased accordingly. In less than three years, the value of the sheep almost tripled – a positive contribution to the local economy. Moreover, the welfare of most farmers involved in the initiative improved significantly since the income generated by selling the sheep was worth more than 1.5 times the capital loaned to them.

In addition to the economic benefits, such as having more sheep in barns and gaining additional income, the initiative also brought about some social benefits to the villagers involved, who gained respect from their families and communities. One of the farmer leaders even managed to send his son to the Bogor Agricultural University, where he obtained his degree in Animal Science last year.

In May 1996, after about five years of continued fieldwork, the initiative officially came to an end. However, since the ewes continued to give birth and farmers kept on returning the sheep offspring, the revolving activities continued on their own.

A New Spur to the Initiative

In early 1999, following the severe Asian financial crises that badly hurt the Indonesian economy, the initiative obtained additional funding to buy 55 new sheep. These sheep were distributed to two other villages in the district of Majalengka.

Meanwhile, around the same time, one former project group in the village of Kadipaten received a 300 million Indonesian rupiah loan through the World Bank's social security network funds. The purpose of this loan was to help the local communities cope with the economic crisis. The farmer group used it to further develop their sheep raising ventures.

Eight years after the initiative terminated, we are still serving the farmers, though at a slower pace and on a smaller scale. To cover the operational costs of supervision and backstopping activities, fieldworkers have been allowed to sell the culled sheep.

Today, livestock production, including sheep production, is growing at a good pace in the district of Majalengka. Both public and private investments have been increasing steadily, especially since the 1998 economic crisis. Between 2002 and 2004, the government of Majalengka invested over 9 million rupiahs in livestock production.

Public funds have been channelled to farmers as loans through local banks, mainly for beef and cattle production. In addition, a big livestock marketing infrastructure has recently been built in the region. Livestock traders from east and central Java bring and sell their sheep and cattle in the newly operated livestock market to local and other traders and buyers from Bandung, Bogor and Jakarta.In other words, the situation of livestock economics in the district of Majalengka has greatly evolved compared to the situation that prevailed in

1989, when the research team first arrived. We believe that over the last 15 years, our work has contributed to this situation by raising a flag to both public and private audiences, informing them about the existing potential in a remote district of West Java that was waiting to be explored and developed.

During recent times, when driving along the river and the irrigation channels in the former research sites, one can see that the irrigation channels are well maintained and that farmers periodically harvest the grass on the riverbanks. From a distance, one can see numerous small white plastic flags stuck orderly in the ground every 10m, on both sides of the irrigation channel. What do these flags mean? Are they a border line? Yes, they are. No one is now allowed to harvest the grass except for the farmer who sticks the flags on the ground. The grass now means something to the rural communities as it has played a key role in local economic development, while contributing to solving environmental problems.

In summary, through participatory development communication, researchers, extension workers, government officials and cooperating small farmers were able to work hand in hand to solve certain community problems.

Through the introduction of sheep raising as an alternative economic activity, the small and landless farmers living in the farming communities along the sides of the Cimanuk River have been able to ensure the management of the meagre public natural resources in their environment.

Problems Encountered

This story would not be complete without describing some of the problems that the researchers encountered in the course of the initiative.

The first problem occurred during the early phase of the fieldwork, when we were trying to gain acceptance and support from the village heads. Most of the village heads in the research sites had shown great interest in the benefits that they could obtain from the activities that were to occur in their areas, and they were keen on having a cut. Denying this opportunity to the village heads could have meant troubles ahead or, worse, it could have led to the failure of the initiative altogether. In contrast, letting them have a share of the benefits could possibly spare the researchers and the farmer leaders many feuds and tensions with the village officials.

For those reasons, the wisest choice at that time seemed to be to give them a small slice of the benefits. With that in mind, we included the village heads' relatives on the list of sheep recipients, provided that they agreed to abide by the project's policies and rules. It seemed to us that by doing this, we would save a lot of energy and avoid further conflict between farmer leaders and village heads. This compromise would also make it possible for us to use the village halls for training meetings.

As the activities unfolded, we found out that high mortality rates occurred among the project sheep raised by the village heads' relatives. Why? The answer was simple. They were not farmers and did not have the necessary skills to raise sheep. Moreover, they did not spend enough time on their tasks. Consequently, they failed in raising the sheep and also failed in returning the sheep offspring.

The second problem we encountered was the implementation of inappropriate feeding practices. Farmers were used to feeding natural grass to their animals and gave similar feeds to the new prolific sheep, without adding enough legumes in the rations. As a result, many animals suffered malnutrition.

Malnutrition brought about many miscarriages among the pregnant ewes and high mortality rates among newly born lambs. To solve these problems, the research team introduced urea molasses block as a protein and mineral supplement for the sheep.

The third problem was the lack of grass and forage supply during the dry season. This recurring problem was seasonal and continuously haunted the farmers year after year. At the beginning of the fieldwork, the research team, together with certain farmer leaders, initiated a demonstration plot to cultivate improved grass and legume varieties in the four research sites. However, very few farmers were willing to cultivate the grass and the legumes, arguing that the grass supplies had always been abundant. Why bother cultivating it?

Another difficulty encountered was that farmers did not respond well to the suggestion of silage-making as a way of preserving the grass, particularly during the rainy season, when there is a large supply of grass. They did not like its smell.

Lastly, after the project's end, when we had to reduce the frequency of our visits to farmers due to limited funds, many farmers did not keep their promise to return good-quality sheep offspring to the initiative. This situation badly affected the efforts to revolve the sheep offspring to other potential farmers. The sustainability question then came up. How much longer would the project last?

Despite these shortcomings, this initiative provided a number of learning opportunities. Looking back on the work accomplished, we can draw lessons and identify the most outstanding features that contributed to making it a success story.

First, conducting a comprehensive baseline study on the research sites was very useful in order to understand the situation, the problems and the opportunities related to local natural resource development, as well as the traditional systems and indigenous capacities of local communities. This also allowed for the development of a sound action plan to implement the initiative.

Second, the strong commitment of a funding partner to ensure adequate funding in order to implement the action plan and prepare the communication materials proved to be a key issue. Furthermore, support from local policy-makers, village authorities and farmer leaders also proved extremely important to ensure the successful implementation of the action plan. The direct involvement of local extension workers and farmer leaders in disseminating the materials, in distributing the sheep and in providing backstopping activities also seems to have greatly contributed to building the necessary trust with local farmers. Moreover, holding monthly meetings proved to be a good way of getting feedback and monitoring activities, while providing further information to farmer groups about sheep raising and marketing their products.

Finally, holding dissemination seminars on the project's success, with the participation of farmer leaders, the academic community and policy-makers, was very important for the project's replication and expansion.

COMMUNICATION TOOLS IN THE HANDS OF FARMERS

Participatory development communication (PDC) aims to establish two-way horizontal communication processes. However, in a situation where there is a need to introduce new information, communication processes can hardly be horizontal if the tools used to communicate with farmers remain in the hands of experts and professionals. This is what the research team of the National Banana Research Project in Uganda had in mind when it established a farmer-to-farmer training programme. Indeed, farmers were so empowered by the process and enthusiastic about sharing their new knowledge with other farmers that they soon wanted to produce their own communication material. After pushing aside the material that had been initially produced for them, they assessed their communication needs, established objectives, defined activities, produced their own material and even selected indicators to measure their success. In other words, they took control of the communication process from beginning to end.

Quiet reigned inside the sub-county headquarters building as farmers of Ddwaniro, within the south-western district of Rakai in Uganda, looked with apprehension at Nora and Moses, the research team's communication resource staff. Nora and Moses fidgeted with the television screen and video deck while Fred, the driver, fixed the power generator. As farmers peeked at the TV screen, they all seemed to be saying the same thing in their hearts: 'Let nothing go wrong now with this video. This is our chance to show other farmers what we have learned after all these months of learning and practising.'

The loud blast from the electric gadgets signalled that all was well and ready with the machines. Moses turned down the television volume as Nora began:

We can now settle down and watch what we recorded last time. Remember, this is the video we are going to use to share knowledge about proper banana management with other farmers. We are nearing the launching day of our farmer-to-farmer teaching programme when we shall invite other farmers, district leaders, researchers and all members of our community and neighbouring communities to show them the results of our endeavours.

The farmers smiled at the thought of moving from the production stage onto the next stage when they would show other farmers what they had been doing during these past months. The video began rolling. Farmers were visibly excited as they recognized themselves in the film:

That, indeed, is Mr Kubo talking. We saw him doing that... That is, indeed, his well-kept banana garden... But why is he talking from far away? It would be better if he were talking while standing near us [in the foreground]... Moreover, he is not looking at us from the screen... The pictures are not flowing well. We should not have used Mr Muganda to illustrate the mulching technology. We could have used Mrs Muganda instead... She is better at explaining issues. But where is the good banana bunch to show the result of a well-kept banana garden? The video continued to a stop just in time as numerous hands shot up simultaneously. Mr Kubo began the barrage of remarks that followed the viewing of the video clip: 'Madame, we have seen the video; but I for myself am not convinced that it will deliver our message.' The farmers unanimously rejected the video, agreeing with Mr Kubo and adding that they could produce a better video than that, but that this time they were going to make a more thorough plan of action. This marked the end of the farmers' meeting with Nora and Moses. The farmers immediately reconvened their own meeting and chose a chairman for the session. The agenda was how to produce a better video than the one they had just watched.

The farmers chose Mr Sebulime to be the overall presenter of their farmer-to-farmer teaching video. Mr Kubo was selected to demonstrate organic manure-making; Mr Lubwamira would demonstrate the digging of trenches in order to guard against soil erosion; and Mrs Muganda would explain the proper mulching of a banana garden. They fixed a date for the next video recording session and informed Nora and Moses about their decision.

On the day preceding the video recording, Nora and Moses arrived in Ddwaniro, together with a professional cameraman. They made contact with the farmers, who took them around the banana gardens that were going to be used for the video illustrations. A mini rehearsal was organized to determine what was going to take place the next day.

Farmers took the lead on recording day. They guided the team of researchers and the cameraman to the different banana gardens that were going to be used for demonstration purposes. They had prepared so well that they knew exactly what sequence the recording would follow. Consequently, the recording process took only a short time, and there was little footage to cut out during the editing of the video. Although this last phase took place in Kampala, away from the farmers, they unanimously accepted the video the next time that it was shown to them. They agreed that other farmers would understand and probably make use of the message contained in the video.

This was confirmed later when they showed the video to other farmers of Ddwaniro.

USING PHOTOGRAPHY

Slowly shaking his head with a sneer, Mr Sebulime examined the photographs that Moses had brought back from Kampala. 'These pictures cannot do. Look, the women appear as if they are going to a wedding feast. How can a farmer working in a garden dress up smartly like this?'

'That is exactly what was recorded last time', Moses replied. 'But it is not right,' Mr Sebulime retorted. 'Look at Mr Bazanya. He is looking directly into the camera; he is actually posing for the photograph. Look, this photograph is so crowded with people that its educative intention is completely lost. No, we cannot use this photograph to teach other farmers. A good photograph should have few people, at least not more than three, and the pictures should be big enough to be seen', Mr Sebulime confidently asserted.

From the side, Nora, who had only recently joined the participatory development communication research team, looked on in disbelief. The farmer was describing a good photograph as if he had attended a photography class. He was only missing the accompanying jargon of 'centre of visual interest' and 'foreground'.

'They always change what they last said', Moses explained. 'We have spent many months on these photographs and sometimes I wonder whether the production process will ever end. It seems as if we are only going round in circles. Farmers photograph what they wish to appear on their brochures; you take the film to Kampala for developing and printing and by the time you bring the photographs back, the farmers have changed their mind and they would like another type of shot.'

In another corner of the same room, Enock, the agricultural extension worker, looked on as the farmer described the picture he had wanted to capture in the photograph. The group of farmers also listened as Bazanya narrated how a good banana crop should be illustrated:

There should be a mother plant with a good banana bunch, a daughter who will take over from the mother, and a granddaughter. The picture should show good mulching practice where you do not put the grass right up to the banana plants. If you put the mulch very near the banana stem, the insects living in the grass will be able to attack the banana plants also.

Listening to the dialogue, Nora realized that the farmer was doing the extension work instead of Enock, who was only listening. At the end of Bazanya's description, Enock asked him why he had not photographed exactly that. In fact, Bazanya had photographed what he had explained. But when the pictures were brought back for the farmers to choose from, the person who had appeared in that particular photograph had taken it for keeps! So it was no longer available for use in the brochure.

MAKING POSTERS AND BROCHURES

Farmers turned up to make corrections on brochures and posters. It was a good gathering considering the long distances, the difficult terrain and the heavy dust on the roads due to the dry season. Although the farmers had previously been divided into three groups according to their different natural resource management problems, work on the brochures went smoothly, with farmers sharing photos when necessary. For example, if a group's photos did not depict what the farmers wanted to illustrate, the group approached other members for a more suitable photograph. This indicated that the farmers appreciated the fact that the exercise was not intended for competition amongst themselves, but for the purpose of sharing information with other farmers who had not yet taken up the new farming practices. It also suggested that the three initial working groups were slowly merging into one group.

The poster was not as easily constructed as the brochure, partly because farmers said that they did not know what posters look like. This was a challenge considering the fact that the community hall in which we were working had a poster on its wall. There were also several posters in the entrance of the community hall. Upon asking the farmers whether they had seen them, some said no, while others said yes – but that they had not examined them closely enough to know what they were about. When examining one poster that depicted proper water and sanitation practices, some farmers actually said that its aim was to teach people how to write. The concept of poster-making was difficult for the farmers to understand since they were expected to put several photos on one chart, which together would tell a story. They finally agreed to make posters with a limited number of pictures. Initially, the plan was to use still photography, with subjects that the farmers themselves photographed. But certain illustrations proved difficult to capture, such as a hole in the ground that was to be used in the making of organic manure and the trenches that capture soil erosion. In the end, farmers settled for artistic illustrations.

During the material production stage, despite time constraints, the production process appeared to have no end in sight. Facilitators then made it possible for farmers to work with a professional video cameraman and an illustrator in order to finalize the process. This arrangement also solved the problem of farmers taking the photographs for keeps and the limited expertise in the farmer-researcher team regarding the making of posters and brochures.

After the production stage, farmers devised a plan of how they were going to use the communication materials. In a workshop setting, farmers agreed on the geographical scope of their intended farmer-to-farmer information sharing. They confessed that, on their own, they could only manage to share information within their villages. This is because farmers can either walk on foot or ride bicycles to cover the relatively short distances in the villages. They also agreed that in order to be credible to other farmers, they had to make sure that their own banana gardens illustrated the recommended soil management techniques.

They identified the different categories of people who needed to be informed about the three modes of soil and water management. These included fellow farmers like themselves, vulnerable groups such as women, as well as the disabled and orphans heading households. In order to get support for their farmer-to-farmer information sharing, farmers included local leaders, politicians and non-governmental organizations (NGOs). They identified institutions and channels through which they could share the information with other farmers. Farmers agreed to make use of radio, a farmers' newsletter and drama to supplement their information sharing through brochures and posters. In addition to one-to-one information sharing, farmers identified village meetings, churches and market days as possible fora. As a way of concretizing their plan of action, they made a time framework within which to accomplish specific activities related to the information-sharing objective. They also agreed on monitoring indicators and mechanisms.

Getting more Specific

As we witnessed this participatory material-making process, we realized that there is another possibility regarding material production for the purpose of farmer-to-farmer information sharing. Three sets of communication materials may be produced. One set is intended for the 'teacher farmers' – the farmers who are going to share information with others. This set of communication materials should be illustrative and should enable interaction between the farmers on the identified subject. The teacher farmers should use this illustrative material to explain to the other farmer(s) the recommended soil management practices. This communication material should be produced in fewer numbers and should be durable since it is going to be used over and over again.

After sharing information with the other farmers, the teacher farmer should leave some material with the learner farmer for reference purposes. This communication material should be produced in larger numbers and need not be as durable as the material for the teacher farmers.

There could also be a third set of communication materials that would act as a back-up for the information-sharing activity by providing brief, general information on the subject at hand. For example, this could be a poster that would hang on the walls of farmers' houses, in shops or in churches; it could also be a radio programme.

In the end, it is important to remember that these materials will always be more effective and appropriate if farmers are closely associated with their production or, even better, if they end up taking charge of their production.

12

Creating Space in Local Forest Management

COMMUNITY-BASED FOREST MANAGEMENT

When, after decades of dispossession and disempowerment, shifting government policies make it possible for a devolution process to take place, local people's organizations can only rejoice. But assuming full responsibility for the environment they live in requires that communities acquire new skills and new capacities, while building their social capital. Although not a panacea, participatory development communication (PDC) can greatly help communities to take up the new challenges associated with decision-making and self-determination.

The Bayagong Association for Community Development Inc (BACDI) is an upland people's organization in Aritao, Nueva Vizcaya, the Philippines.

Its members were victims of the involuntary relocation project brought about by the government's construction of a dam that submerged the people's ancestral lands. Instead of taking the government's offer of resettlement in a place not akin to their culture, they migrated to a nearby forested land. This experience precipitated their engagement in some kind of participatory development communication as they needed, then, to discuss their fate and how they could cope with it. Of course, these people did not know that they were involved in PDC!

Labelled as squatters and encroachers by government, they struggled for almost 28 years to hold on to the lands that they have been *de facto* occupying since they migrated in 1960. Periodically during the past, they had to resist and endure the efforts of government agents and threats from outside speculators to expel them from the area. To make their appeals known to the concerned authorities, these upland farmers engaged in dialogues with government officials and authorities. But when their voice seemed not to be heard, they resorted to street protests and lobbying. These strategies were outcomes of their frequent discussions and analysis of their situation. A wind of change worked in BACDI's favour when the government's policy on natural resource management adopted community-based forest management as the national strategy for sustainable development of open forestlands. BACDI applied for and was awarded the Community Forest Stewardship Agreement, a tenurial instrument that formally legitimized their claim over the forestland they have been occupying. It granted them tenure of 25 years, renewable for another 25 years.

While the political and social intent of such a devolution strategy was highly appreciated, issues emerged regarding the community's readiness and absorptive capacity to handle the devolved tasks and functions. Devolution as used here refers to 'the process where the locus of power and control shifts from the state to the local communities'. Under community-based forest management and as part of their new set of responsibilities, BACDI members underwent the process of participatory resource management planning. Here, PDC took the form of a social preparation method, equipping the people with the knowledge and skills that they would need as forest resource managers. Through community mapping, resource identification and problem prioritization, they learned to draft their plan and called it their own.

The other gender-sensitive participatory rapid appraisal methodologies used for situational analysis included community resource profiling; gender-disaggregated household activity and decision-making; stakeholder identification and analysis; political-ecological mapping; historical-structural analysis; institutional analysis; participatory analysis of problems and options; and resource management action planning. In all these methods, PDC was a core process for learning and planning.

Supplementing this variety of PDC methods were communication tools such as Venn diagrams, maps, community billboards and posters, written documents of their organization, and policies agreed upon for managing their resources.

This exercise paved the way for the community members to have a better grasp of the quality of their resources, their cultural integrity, their social capital and their political capacity, as well as a sense of their prejudices, weaknesses and vulnerabilities – knowledge that they could have not unravelled had they not undergone PDC. This also enabled them to identify internal and external threats to their resource management and how these might be handled.

To a large extent, the participatory experience enabled them to implement their plan with more confidence and better direction. Whereas before, the community simply remained buried in their culture of silence and subservience, the knowledge acquired of their rights and responsibilities through community meetings and discussions enabled them to become more open and assertive. As a result, when they conducted social mobilization among their members and allies, they were already more aware of community organizing, government policies, their internal capacities, external possibilities and the given constraints in the field.

Participatory development communication paved the way for BACDI's link-up with other government and non-governmental organizations (NGOs). Members knew that they could not muster all the resources needed to manage their forest resources well. They eventually realized that they also needed to access external assistance, such as livelihood opportunities, *barangay* roads, markets, credit, schools, health centres and other social services. Partnerships became another imperative, and they were able to establish alliances by using PDC.

As they learned more about themselves and their resources from the various participatory rural appraisal methods used in the study, BACDI members were able to come up with more rational plans and approaches for managing their communal forest. In the past, they had left these matters to their leader and whoever was deemed influential in their group. PDC also enabled them to reorient their outlook and address their needs by first using locally available resources before turning to outside sources for assistance.

For monitoring and evaluation, they employed informal, unstructured discussions where they would ask each other and reflect upon where they were at a certain time relative to their plans and targets. The observations they made were then discussed in their regular community meetings. Here, members were allowed to clarify and validate their observations.

Used in the various facets of participatory resource planning, PDC has actually served as a mechanism and, at the same time, as the context of group learning. As they progressed in learning about their biophysical, socio-cultural, economic and political environments, BACDI members became more enlightened and rational managers of their forest resource. Even though most of them had only attained a low level of education, and some were even unschooled, they felt that they had learned many things from their community natural resource management activities. As a result, they were able to delineate the bounds and limits of the physical and political space that they had claimed for themselves. This was reflected by the sample plan and set of policies formulated governing the sustainable use and protection of their resources.

Social dialogue became a frequent activity in which they engaged by virtue of the communal nature of their forest resource. Decisions had to be collective and inclusive. Although discussions were not always smooth sailing, the members of the community have gradually learned the techniques of negotiation and consensus-building. In the process, their dwindling social capital has been enhanced. Social capital refers to cooperative social relations and collective action processes. This is embedded in 'norms or reciprocity, networks of civic engagement, trust, and obligations that facilitate coordinated activities' (Asian Development Bank, 1994). Since social capital depreciates through time, constant dialogue and open communication prevented it from deteriorating. Strong social capital has enabled the BACDI community to protect their traditional resource system from outside encroachers and the centralizing tendency of the state.

As BACDI members learned to master the science and art of forest management, they were continuously bombarded with challenges. Dominant among these was the multi-stakeholder setting for decision-making and action planning. It was through PDC that community members were able to respect and manage the diversity of views on an issue. Constant dialogue opened them to many possibilities and made them realize that there could, indeed, be many possible solutions to a problem. Likewise, PDC provided the venue for mainstreaming women in a men-dominated BACDI ethnic group. From mere cooks and servers during community meetings, women started assuming more substantive roles, such as being a liaison officer, secretary, treasurer and even *barangay* councillor. It must be emphasized that it was not PDC alone that paved the way to local forest management by BACDI members. Other factors were also involved, such as forest culture, social capital, policy presence and assistance by external actors. It can be safely assumed, however, that participatory development communication played a critical role, tempering the socio-political environment, internal and external to the community, so that a climate favourable to the community's takeover of resource management was created.

REFLECTIONS

Just like any other communication methods, PDC is not a panacea. It also has certain caveats, nuances and limitations. As a tool, its users and practitioners have to understand its profound intricacies. It takes time and practice, and a lot of learning from failures, to appreciate what it can and cannot do. Hence, its participants have to understand the action–reflection–action dynamics that are built into it. Participatory development communication, as a catalyst for change, should also be accompanied by the inputs necessary for development, such as credit or capital, roads, water and technical assistance. It can establish the link between and among the development players in the community. Information and consensus are necessary conditions; but they are not sufficient to bring about broad development as desired by the community.

Likewise, devolution does not necessarily ensure effective local forest management. Policy and political and technical support have to be provided. But the use of PDC as an ingredient in the process makes it more workable, socially acceptable and satisfying in meeting the democratization and efficiency agenda of natural resource management. Moreover, PDC enables the evolution of 'participation-as-engagement' process, veering away from the usual 'participation-as-involvement' process. The latter simply implies that the 'subject' or actor is merely a participant in the process, in a position of powerlessness. The former, instead, transforms the actor into an 'active subject' and not just an object or a client. 'Participation as engagement' is a necessary condition for genuine empowerment and for creating space in local forest management, which can be best achieved through participatory development communication. The impact on the community, especially in terms of managing their own forest, would have been different had other forms of top-down communication been employed.

FROM INFORMATION TO COMMUNICATION

In Burkina Faso, as everywhere in Africa, radio exerts an undeniable attraction. In countries where the government has given up its monopoly of the airwaves, nearly all the available frequencies are now occupied by radio stations run by community, commercial or religious broadcasters. Not only do people everywhere listen to the radio, but it has taken root throughout the land, even in the most remote villages. The success of this tool, which some call 'Africa's internet', allows people to express their concerns in their own languages. Yet, this communication medium remains largely an unexploited resource when it comes to participatory development communication (PDC). An experiment conducted by the Journalists in Africa for Development (JADE) network at three places in Burkina Faso and in Mali demonstrates both the usefulness of such an approach and the importance of taking certain precautions to ensure its success.

Somewhere in Burkina Faso a farmer sits down in a studio and belts out a few songs dedicated to his friends and relatives. The man is about 50 years old. He is happy to be able to make his voice heard in his home village. This is the first time that he has ever entertained a radio audience. Earlier, he had sent his sons with a present of some yams and a chicken for the station technicians.

Elsewhere, members of a rural listeners' club have just delivered their weekly programme sheet to the station. The form is filled out in the national language and shows the broadcasts that they have heard, together with listeners' observations and their requests. The station will take these into account in its coming broadcasts.

Examples of this kind represent amazing progress compared with the way things were once done, from the time the first radio stations were established in Africa until the 1990s, when many countries in West Africa began to deregulate the airwaves. Yet, despite this freedom of speech and the openness that radio maintains with its audience, considerable effort is still required if radio stations are to become true tools for participatory communication. It is all very well to free up the airwaves or to run radio games in the villages; to greet people throughout the day by their own names; to produce and broadcast programmes about farming, livestock and the environment; and to take account of listeners' expectations. All this allows better use to be made of radio;

but it is not the same thing as making radio a tool for PDC, and still less for the management of natural resources.

There is no magic recipe for making radio a participatory communication tool. But the experiment that JADE conducted during 2000 to 2002 in three places (two of them in Burkina Faso and one in Mali) provides some food for thought. This chapter looks, in particular, at the most successful effort, launched in January 2001 at Ziniaré in central Burkina Faso.

Central Burkina Faso is a region of poor soils and one that faces severe water shortages. The 2001 project involved using participatory communication to share information among communities, development agents and radio producers in order to improve the management of natural resources. Besides JADE, the group includes technical experts from the agriculture and environment ministries, farmers organized as communication relay points in their communities (also called local communicators) and radio producers. The partner radio station belongs to a farmers' organization, Wend Yam.

TRADITIONAL CHANNELS: LOCAL COMMUNICATORS

The project on rural communication and sustainable development made some innovations in natural resource management by encouraging reliance on local skills, particularly local communicators. The local communicator was the cornerstone of the project and served as the interface between the relay points (*i.e.* the project representative in the area) and the community. The local communicator's many functions make him an indispensable player in any action-research effort.

In fact, local communicators perform many tasks: organizing and hosting chat sessions and supporting the research team in identifying natural resource management issues with the community. They also gather local knowledge to be dealt with by the broadcast, identify the resource persons (model farmers), organize recording sessions and gather feedback.

THE IMPORTANCE OF FOLLOW-UP

The project's impacts are due, in large part, to follow-up by stakeholders in the field. Development technicians have taken it upon themselves to share experiences, experiments and lessons learned as a way of working with the communities. They can now foster synergy among themselves and, in particular, with the radio producers. As to the radio producers, they can regularly be found collecting information in the field. They also work with the community in choosing issues, in handling them and in gathering feedback. For this purpose, they have retained the devices that were introduced (*e.g.* listeners' clubs and local communicators). Today, radio is using the participatory communication approach to go into the field, identify problems and seek solutions with local people. Moreover, thanks to local communicators who have their roots in the community, the issues of concern to local people can be inventoried and, in this way, can form the basis for radio programmes. Local people are now taking a much more active role in producing broadcasts.

In Sikasso, radio producers who were not involved in the project were, nonetheless, impressed by its achievements to the point where they wanted to use the approach in their own work. Within government technical services, as well, several managers recognized the usefulness of radio in dealing with bushfire prevention, sharing knowledge about production techniques, settling disputes over environmental management, preventing disease and reinforcing producers' groups.

Today, Radio Yam Vénégré has more direct and more regular contact with local people and is facilitating negotiation between them and the development agencies. With its experience in PDC, it is now playing an advisory role to development institutions. Moreover, it has helped a project that was designed to strengthen farmers' organizations to clearly explain the objectives of its surveys before beginning fieldwork. Programmes have been produced and broadcast for this purpose, and development officials are now working together. Even outside the project, broadcasts are being produced with other agencies attracted by the approach, something that also earns new revenues for the radio station.

By way of conclusion, we can take satisfaction from having used radio in a new way and we can congratulate ourselves on the resulting achievements. But the research team could have gone even further. It should have systematically produced audio cassettes in order to preserve the chain of communication, to reach those who missed the broadcasts and to allow certain points of debate to be revisited.

Posters or booklets would also have helped to reinforce viewpoints. With respect to institutional information – that is, community questions about the project itself and the responses of the project coordinators – use of the radio could have cleared up a number of misunderstandings. This experiment provided a good opportunity for assessing the most efficient and effective types of programming. In addition, PDC could have been compared with the conventional communication format, which is simply based on broadcasting information. Recognizing that the broadcasts occurred both before and after the group discussions, a summary of the previous broadcast could have been replayed before the discussion sessions were held. The radio station, in fact, had the right equipment, including powerful loudspeakers, to perform this task properly.

Despite the importance of rural radio, it should not take the place of chat sessions. Radio should be seen as a supplement, one that relies on the results of these sessions for input. Chat sessions must also meet certain conditions if they are to be truly participatory. In the absence of all these conditions, people will, of course, continue to listen to the radio; but what they hear will be the voices of the producers, the dominant voices of the day – it will not be a tool at the service of the community that is truly participatory.

NATURAL RESOURCE MANAGEMENT

Radio is usually seen as the ideal medium to reach people in remote areas who do not have access to other sources of information. But when committed to community participation and people's development aspirations, radio can be much more than a source of information. It can also be a powerful tool to facilitate consensus-building and decision-making at the community level, thus becoming a catalyst for change. This is precisely what Radio Ada set out to do when it decided to accompany the efforts of the people of Obane in the eastern part of Ghana. After four decades of watching their environment deteriorate, the inhabitants of that community rolled up their sleeves to restore the waterway, which used to be at the heart of their lifestyle, and regain their lost prosperity. Day in, day out, Radio Ada was by their side, voicing their concerns and acting as a morale booster. In the hands of the community, the radio also became a tool for advocacy and mobilization. The results are quite astonishing.

A Catalyst for Collective Action

Many towns and villages seem to have forgotten the reasons that they have come to be where they are. The people of Obane have not.

Obane is a rural community in Big Ada in the Dangme East district of Ghana, about 100km from the capital city of Accra. It is a brisk hour-long walk away on a hot, dusty road from Radio Ada, the community radio station of the Dangme-speaking people.

The people of Obane still remember that the waterway is the reason that their forefathers chose to live there. The waterway, the Luhue River, is a tributary of the mighty Volta River. It supplied fish, provided water for irrigation and served as a bustling transport course. The water was so abundant that Obane used to be the food basket for Big Ada. The women were, among other occupations, fishmongers, farmers, mat weavers and petty traders, while the men fished, farmed or hunted. That was over 40 years ago.

The creation of the Volta Dam during the early 1960s reduced the flooding of the fields at Obane. Weeds, trees and debris choked the waterway. The surrounding lands became barren, leaving a fetish grove with isolated trees and some patches of green to the south. Not even the shadowy line of emaciated trees tracing the meander of the Luhue River in the background is thick enough to break the line of sight to the west.

Over time, Obane became one of the poorest communities in the district. For this reason, Radio Ada has always taken a special interest in Obane. Radio Ada is Ghana's first community radio station. On the air since February 1998, the station broadcasts 17 hours daily exclusively in Dangme, the language of its listening community.[1] It is staffed by volunteers drawn from the community and trained in its home-grown workshops. Radio Ada's identity is rooted in the culture of its listening community and inspired by their desire to improve their economic way of life while maintaining close community ties.

The station actively pursues a participatory development philosophy. One of the main ways in which it tries to operationalize this philosophy is through 'narrowcast' programmes. These regular weekly programmes are recorded in different communities, with the main occupational groups in the listening community – fishmongers, fishermen, women farmers, men farmers and so on. The programmes are driven by the participating community members. They determine the content and it is their voices that predominate. Programmes are presented as a continuing dialogue, where producers simply act as facilitators. The station's holistic approach to community-initiated development as 'the voice of the voiceless' in content and the production process generates trust.

Thus, the programmes serve to affirm the knowledge and experience of group members, who share what they know, what they feel and what they believe. They fuel consultation and interaction between themselves and their counterparts, as well as between other listeners in the Dangme-speaking communities. They also serve to deepen the relationship with Radio Ada. As resources permit, the stations follow up the programmes with more extensive community consultation. It was after a number of 'narrowcast' programmes had been produced with the people of Obane that one such consultation was initiated by Radio Ada. The consultation was held a few years ago, when the community was confronted with a decision to migrate or to renew their waterway.

The consultations generally involve the use of participatory rural appraisal (PRA) tools, such as the ranking-and-scoring matrix for needs or priorities. Despite the inherent friendliness of the tools, often when one asks communities about their priorities, the answer goes something like this: 'We need pipe-borne water, roads to transport our produce, electricity, employment, etc.' The litany never seems to end and tends to read like a shopping list.

In the case of Obane, possibly because of the ease and trust that had been built in their relationship with Radio Ada and almost certainly because they were at a decisive point in the life of their community, the responses were more detailed, reflective and textured. In every case, respondents had a story to tell connecting the present to the past.

The fishermen lamented that they had become goat and sheep rearers; one owns a cattle kraal. They were worried that the different fishing skills they had learned from their parents and friends were no longer of value to them and their children. The river did not exist during most of the year. It was choked, and even during the short rainy season it provided only limited access. The vast marshy lands that had boasted of crabs and reeds for weaving had all run dry. The women could not bear sending their children off to school in Big Ada and beyond without enough food to support their stay. Formerly a little rise in the tide caused natural flooding of the fields for fresh crops even during the dry season. Although Obane lacks many basic amenities, the women identified 'unity' as their most important priority for development. They were sure that they, working with their husbands and joining other neighbouring communities, could revive the economic life of the community and restore it to its former status as a food basket.

During the plenary discussion, Divisional Chief Nene Okumo was invited to share his reflections. He narrated how during the past their fathers occasionally cleared the waterway to support ecological processes and to sustain flora and fauna. Then and there, the plenary decided to dredge the Luhue River using communal labour. Immediately, Madame Adjoyo Djangma, a mother of seven and redundant fishmonger, intoned a jubilant, melodious traditional chorus. The response was infectious as others took up the chant. There were contour lines on some foreheads – more than 40 years of vegetable growth and silt to clear! The work extended over 10km, and would entail varying degrees of difficulty. In the meanwhile, the other neighbouring communities had their own priorities and there was the need for tools and other logistics. How could they possibly mobilize the resources needed?

'How can your community radio station, Radio Ada, help?' There was a measure of relief on some faces as the role of communication was put forward: 'Announce what we have decided to do... announce the day and time for the work... when others hear, they will join us... announce the names of those who have reported for work... put our needs on the radio... tell other people about the by-laws we shall make to prevent the river from choking.' The ideas kept coming. Having facilitated consensus- and decision-making on the ground, Radio Ada fuelled the communication bonds within the Obane community and connected them to other communities and institutions to support the communal labour needs. Trusting the station to honour its offer of support, the groups actively turned it into a tool for advocacy and mobilization. They announced at their convenience and free of charge their work plans – whose turn it was to work – and raised issues on air about the environment, their occupations, their lives and the project's progress.

Both the women and the men of Obane conducted the clearing, and 60,000 seedlings of red and white mangrove have since been planted. Before long, four groups from different communities – Obane, Gorm, Togbloku and Tekperkope – were working together to dredge the Luhue River. Other groups from other communities joined them to show solidarity and were duly acknowledged on air. They included communities from Dogo, Dorngwam and Atortorkope, as well as from Aminapa, Wasakuse and Midie.

District Wildlife Officer Dickson Yaw Agyeman stated: 'The radio broadcasts were a morale booster and a challenge. The Wildlife Department gave them Wellington boots, nylon ropes, cutlasses, hoses for weeding and food for work – kenkey and fish and pepper.' Their voices and active participation on air also won them other collaborators, allies and supporters. Organizations and institutes such as the Dangme East District Assembly, the Canadian High Commission, Green Earth and the Kudzragbe Clan of Ada Elders contributed to the initiative. The men's group received a loan facility of 14 million Ghanaian cedis to boost agriculture, while the women obtained a loan for 15 million cedis for food processing and marketing. In addition, a 12-seater KVIP toilet was built. Funding was through the Wetlands Management and Ramsar Sites Project, which is supported by the World Bank. The magnitude of work accomplished over these four years might intimidate a stranger; but knowing what the river and the surrounding fields mean to their lives, the people of Obane persisted. With the support of other communities and of the Wildlife Department, as well as their community radio station, Radio Ada, they now have what matters most to them.

Day in, day out, as needed, Radio Ada followed the efforts of Obane on the air. Throughout the process, it took its cues from the leadership of the community, who were constantly at the station making requests for specific programmes and announcements. Radio Ada not only helped to mobilize concrete material support for the back-breaking work of the people of Obane, but also inspired the rest of the community with their enthusiasm, perseverance and sense of unified purpose. By consistently projecting a community endeavour into the public domain, as well as through the soft cheerleading style of the radio broadcasts, even disputes over land boundaries, customs and leadership were avoided. No one could afford to risk name or reputation by swimming against what was now literally a growing tide. Today, one can travel on a boat from Big Ada through Luhuese to Obane and beyond. Year-round small-scale farming is possible by using irrigation methods, and the people now have access to freshwater. A replication is also in the offing at Totimekope near Ada Foah, the administrative centre that is the twin town of Big Ada. There, the Futue River is to be cleared in order to restore river transport, fishing and farming. The satisfaction of Radio Ada is in fulfilling its mission and sharing in the joy of the community's successes. Occasionally, however, there are more concrete rewards – perks, so to speak.

One sunny day, one of the dredging groups, led by Alfred Osifo-Doe (a professional mason), burst straight from the dredging site, sweating and jubilant, into Radio Ada, excitedly bearing a gift for the station – proof of the success of the common endeavour, they cried, fished out of the increasingly swelling waters of the Luhue. Deeply touched, but mindful of its conservation role (as well as of the practical difficulties involved), the volunteers of Radio Ada managed to persuade the dredging team to return the gift to its natural habitat. So, be careful the next time you cross the Luhue River because by now it must have grown – the baby crocodile, that is!

IDENTIFYING PROBLEMS IN NATURAL RESOURCE MANAGEMENT

'What do you consider to be the priority problem with water, wood, fields and the land, and how could you help to resolve it?' This was the first question put to debate in the four discussion groups, and then in a plenary session. The idea was to highlight the community's capacity to find its own means for resolving its problems before turning to other sources for assistance.

At Nagreongo, an arid region of central Burkina Faso, three issues emerged from the discussions: water shortages, soil degradation and lack of wood. A poll was taken, and water shortages received the most votes. This issue was therefore ranked first in terms of priorities. Among the reasons for the water shortage, people cited the inadequate wells, the great depth of the water table, the lack of erosion control and the absence of reservoirs. A more thorough analysis also revealed that the water was unhealthy for human consumption, and that simple techniques would have to be found to make it drinkable.

In terms of consequences and impacts, it was evident that the lack of clean drinking water led to public health problems, such as Guinea worm disease, childhood diarrhoea and other waterborne illnesses. Moreover, the shortage of water was destroying village communities as their able-bodied workforce gradually left, and was also harming livestock activities.

PROGRAMMING PRODUCED BY RADIO PERSONNEL AND LOCAL COMMUNICATORS

In a participatory approach, producing a magazine-type programme involves not only radio producers, but also local communicators. These are farmers who are responsible for identifying village interests in advance, preparing field trips, taking part in producing a broadcast and gathering feedback from listeners. In this way, a series of 45-minute radio magazine programmes was produced, essentially based on ideas from members of the various discussion groups and interspersed with a few musical interludes. This was 'raw feed' that served to highlight participants' viewpoints.

Broadcasting the Programmes

Public announcements were made, usually a week in advance, alerting people of the coming broadcast. This not only made it possible for people to tune in individually, but also, in some cases, to form discussion groups.

Gathering Feedback

The local communicators were responsible for sounding out listeners' views on the broadcasts, with the assistance of the listeners' clubs (in Ziniaré and Sikasso). Typically, a local communicator would visit the main regional town on market days, and people would pass on their opinions to him or her.

In some cases, listeners with literacy skills would send their viewpoints direct to the station. Generally, the station attempts to respond directly to listeners whenever possible. It also occasionally calls upon resource persons or institutions to deal with listeners' concerns.

Lessons from the Experiment

Preliminary steps are needed to establish the sense of trust that is essential in getting the process up and running. Radio stations typically approach the communities with a preconceived idea of a broadcast. People rarely have the chance to make their expectations known in advance. Even today in rural areas, radio stations, despite their proliferation, are still inaccessible to most of the population. The fact is that when people are able to express themselves over the airwaves, this gives them some power, not only within their village but beyond. A good example of this arose in Nagreongo. Radio producers went to the village for a working session with resource personnel in order to establish a basis for collaboration and they found, to their surprise, that the entire village came out to meet them. Yet, this is a common practice, something communities do to stay in the good graces of development agencies. How could the situation be handled? It was here that local knowledge helped to avoid embarrassment. The working session originally scheduled was set aside, and the programme producers set about recording a broadcast on the history of the village, to the great satisfaction of everyone. They then held an interview with the local healer, an influential personality in the village and a great master of ceremony. These broadcasts were carried over RadioYam Vénégré in the evening.

In this particular case, the radio team turned a dicey situation into a triumph by producing a radio broadcast on the spot. This ability to react quickly to situations allowed them to win the villagers' trust, and then to hold discussions with separate groups of men, women, youth and the elderly.

Collaboration with traditional authorities, technical experts and communities (in other words, teamwork) is an essential factor for success

The participatory approach demands real collaboration among all stakeholders – traditional authorities, research teams and communities – in order to minimize the risk that things will grind to a halt if certain parties are overlooked. Teamwork produces results that reflect the great diversity of the group. In Dori, as in Sikasso, field reporting was considerably enriched by the representatives of farmers' organizations and by rural development experts. The programme producers confined themselves to formulating pertinent questions and ensuring that proper use was made of the recording equipment. Yet, this approach requires a thorough knowledge of the local setting.

RADIO GIVES ADDED VALUE TO LOCAL FOLK KNOWLEDGE

The importance of local folk knowledge has to do with the fact that it is slowly disappearing in the areas under study, without ever having been fully assessed and appreciated. Folk knowledge is not a simple store of information; rather, it has to do with the realm of understanding and of what is sacred. This is why it is important that communities are involved and that the appropriate approach, involving the following steps, is taken:

1. Pinpoint and prioritize the natural resource management problem.
2. Identify the resource persons who have local know-how relating to that problem.
3. Interview those resource persons.

The information collected must deal with the nature of local knowledge, where it was learned, its description, the way in which its holder uses it in practice, the results obtained and the constraints observed. It is also useful to develop in advance a line of argument to support any information-gathering process in order to overcome the possible reluctance of local knowledge holders and to appreciate their knowledge, which is generally brushed aside by development officers.

That line of argument might point to the fact that as the processors of this folk knowledge move on in years, it becomes ever more urgent to record and preserve what they know and to make it available to the entire population. It may also be useful to point out the paradox inherent in looking to outside knowledge, which is often costly and inappropriate, when there are readymade home-grown solutions at hand. Local knowledge, indeed, must be treated as a heritage to be preserved.

How can radio help to promote traditional local knowledge? Because it is immensely attractive to the public, radio can give to any information that it broadcasts an importance that it would not otherwise have, not only in terms of building an audience, but above all in terms of winning recognition for it. To be quoted on the radio leads to increased credibility.

PARTICIPATORY COMMUNICATION: THE KEY TO NATURAL RESOURCE MANAGEMENT

Natural resources are, by their nature, common resources, and their exploitation and (above all) their preservation are in the hands of the community. These resources must be regarded as a fragile asset to be used wisely while being conserved for tomorrow, and protecting them depends in large part upon the community. It is the community who must decide whether or not to preserve forests, water sources and the entire natural environment in order to ensure its own subsistence and that of future generations. Yet, the general tendency is to exploit these resources with a view only to the short term, and there is little thought given to preserving them. We may safely say, then, that there is cutthroat competition between different users (farmers, herders, agro-businesses, etc.) that risks not only mortgaging the future of these common resources, but also creating conflicts over access to them. For all these reasons, natural resource management is a natural target for participatory communication as a means of creating dialogue among all stakeholders and helping them to make decisions that will serve the interests of everyone. Above all, PDC can help to organize joint activities based on a shared vision.

PROMOTING SOCIAL GROUPS THROUGH RADIO

It was not by chance that JADE chose to work with radio: it already had long experience in this field. But the first attempts, innovative though they were, did not meet with much success for two reasons. First, JADE was broadcasting with the help of technicians, radio producers and local communicators; but it was not working with the communities themselves, who were treated strictly as sources of information. Second, radio was not used to support participatory communication activities that might help to identify problems, analyse them and find solutions.

Moreover, despite its community radio designation, RadioYam Vénégré, JADE's major partner in this project, remained a typical rural radio station. It was content to broadcast information to the rural public, the only difference being that it made much use of Mooré and Fulfuldé, the two most widely spoken languages in the region. The results produced by this top-down approach were disappointing and led JADE and its partner to try to use radio in another way. The new approach recognizes and places value on community viewpoints. The public can express itself directly over the airwaves, instead of simply listening to broadcast advice from experts and radio producers. Discussion groups (for women, young people, adults and seniors), backed up by a team of development experts and radio producers, analyse problems with natural resource management, their causes, their consequences and their potential solutions. This process, which has been recorded from start to finish, involves four stages:

1. Identifying problems in natural resource management;
2. Programming produced by radio personnel and local communicators;
3. Broadcasting the programmes;
4. Gathering feedback.

Index